Disaster Risk Reduction

Methods, Approaches and Practices

Series Editor

Rajib Shaw, Keio University, Fujisawa, Japan

Disaster risk reduction is a process that leads to the safety of communities and nations. After the 2005 World Conference on Disaster Reduction, held in Kobe, Japan, the Hyogo Framework for Action (HFA) was adopted as a framework for risk reduction. The academic research and higher education in disaster risk reduction has made, and continues to make, a gradual shift from pure basic research to applied, implementation-oriented research. More emphasis is being given to multi-stakeholder collaboration and multi-disciplinary research. Emerging university networks in Asia, Europe, Africa, and the Americas have urged process-oriented research in the disaster risk reduction field. With this in mind, this new series will promote the output of action research on disaster risk reduction, which will be useful for a wide range of stakeholders including academicians, professionals, practitioners, and students and researchers in related fields. The series will focus on emerging needs in the risk reduction field, starting from climate change adaptation, urban ecosystem, coastal risk reduction, education for sustainable development, community-based practices, risk communication, and human security, among other areas. Through academic review, this series will encourage young researchers and practitioners to analyze field practices and link them to theory and policies with logic, data, and evidence. In this way, the series will emphasize evidence-based risk reduction methods, approaches, and practices.

Anindita Sarkar · Nairwita Bandyopadhyay ·
Shipra Singh · Ruchi Sachan
Editors

Risk, Uncertainty and Maladaptation to Climate Change

Policy, Practice and Case Studies

Springer

Editors
Anindita Sarkar
Department of Ecology and Natural Resources Management
Center for Development Research (ZEF)
University of Bonn
Bonn, Germany

Department of Geography
Miranda House
University of Delhi
New Delhi, India

Shipra Singh
Department of Geography
Miranda House
University of Delhi
New Delhi, India

Nairwita Bandyopadhyay
Department of Geography
Haringhata Mahavidyalaya
University of Kalyani
Kalyani, West Bengal, India

Ruchi Sachan
Department of Geography
Miranda House
University of Delhi
New Delhi, India

ISSN 2196-4106 ISSN 2196-4114 (electronic)
Disaster Risk Reduction
ISBN 978-981-99-9473-1 ISBN 978-981-99-9474-8 (eBook)
https://doi.org/10.1007/978-981-99-9474-8

This Springer imprint is published by the registered company Springer Nature Singapore Pte Ltd.
The registered company address is: 152 Beach Road, #21-01/04 Gateway East, Singapore 189721, Singapore

Foreword

With impacts of climate change being faced disproportionately by the most vulnerable who have not contributed to the problem in the first place, the world climate community is increasing, raising calls for more and effective adaptation. However, not all adaptation reduces climate risk, and when adaptation efforts increase risk, it is called maladaptation. Avoiding maladaptation is important, but it is not easy to know what will be maladaptive apriori. "Risk, Uncertainty and Maladaptation to Climate Change—Policy, Practice and Case Studies" is a milestone in our collective journey to understand the multifaceted challenges we face. I commend the editors and contributors of this remarkable book for their vision, their unwavering commitment, and their invaluable contribution to literature and knowledge. Through this anthology, the editors transcend traditional narratives, and offer us more insight on climate risk, uncertainty and maladaptation. The diversity of voices represented here is a reminder that solutions to global issues must be as diverse as the people they aim to serve.

I compliment the editorial team of the book comprising entirely of remarkable women from the Global South to represent a tapestry of experiences, cultures, and backgrounds, united by their commitment to amplify the voices and perspectives, often unheard, in the broader discourse on climate change. This book stands as a testament to the power of various voices in tackling one of the most urgent global challenges of our time. The stories, analyses, and solutions it presents reflect a deep well of wisdom nurtured by countless generations who have harmonized with their environments, and who now confront the consequences of a changing climate with unwavering resolve.

In the pages of this collection, you will encounter perspectives that defy preconceptions and narratives that inspire change. It is array of thought-provoking and appealing contributions from authors of various disciplines, each offering a unique methodology and outlook to the unprecedented impacts and threats of climate change, (mal)adaptation strategies and resilience of communities.

I wholeheartedly applaud the women editors who have compiled this editorial volume to be the book series *Disaster Risk Reduction—Methods, Approaches and Practices*. Through their collective acumen, this book stands as a symbol of hope and a guide to redefining resilience and innovation. May their voices inspire us all to take action, to work collaboratively, and to reshape our world into one that is not only resilient but also equitable, just, and sustainable.

Nairobi, Kenya

Dr. Aditi Mukherji
Director
CGIAR Climate Change Impact Platform
IPCC Author
AR6 cycle

Acknowledgements

The idea of this book emerged from an international webinar on 'Addressing Risk, Uncertainty and Maladaptation to Climate Change' organized by the Department of Geography, Miranda House, University of Delhi. The deliberations made in this webinar by experts from across the globe with a diverse range of expertise on several aspects of climate change, became the basis for conceptualization of this theme. We are indebted to Prof. Rajib Shaw, Professor, Graduate School of Media and Governance, Keio University; Dr. Animesh Kumar, Head, United Nations Office for Disaster Risk Reduction, Bonn, Germany; Dr. Zita Sebesvari, Deputy Director, United Nations University Institute for Environment and Human Security; Prof. Lyla Mehta, Professorial Fellow, Institute of Development Studies, University of Sussex; and Ms. Irina Rafliana, National Agency for Research and Innovation, Indonesia for their insights which became an integral part of our book.

We express our heartfelt gratitude to each one of the authors who contributed their time and expertise to this book. The rich diversity of methodologies and multidisciplinary conceptual understanding has added enormous perspective to the knowledge on climate change adaptation studies. Primary surveys to the analysis of satellite imageries have helped to include a vast range of data and perspectives in this book. We would also like to express our appreciation to the contribution of the research participants, local communities, and organisations who have helped with information and data collection for the chapters of the book. Without their support, this book would not have become a reality. We also acknowledge the hard work of the anonymous reviewers for their constructive suggestions that has helped to improve the quality, coherence, and content presentation of chapters of this book.

We are grateful for the incredible support from our colleagues, students, parents, family members and friends for their unwavering support in the most challenging times. Their belief in our work have been a constant source of motivation. We extend our deepest appreciation to all those who have contributed to this book, their support and guidance have been invaluable.

Prof. Dr. Anindita Sarkar
Senior Researcher
Institute of Development Research (ZEF)
University of Bonn
Bonn, Germany

Professor, Miranda House
University of Delhi
New Delhi, India

Dr. Nairwita Bandyopadhyay
Assistant Professor
Haringhata Mahavidyalaya
University of Kalyani
Kalyani, India

Ms. Shipra Singh
Assistant Professor
Miranda House
University of Delhi
New Delhi, India

Dr. Ruchi Sachan
Assistant Professor
Miranda House
University of Delhi
New Delhi, India

Contents

Contributors

Khalid Azim National Institute of Agronomic Research, Regional Center of Agadir, Agadir, Morocco

Sunando Bandyopadhyay Department of Geography, University of Calcutta, West Bengal, Kolkata 700019, India

Tarik Benabdelouahab National Institute of Agronomic Research, Regional Center of Rabat, Rabat, Morocco

Nathoo Bharati Department of Economics, Babasaheb Bhimrao Ambedkar University, Lucknow, India

Anshuman Bhardwaj School of Geosciences, University of Aberdeen, Aberdeen, UK

Phup Kesang Bhutia Department of Geography, Cluny Women's College, Kalimpong, India

Youssef Chebli National Institute of Agronomic Research, Regional Center of Tangier, Tangier, Morocco

Kamal Choudhary Department of Anthropology, University of Delhi, Delhi, India

Fouad Elame National Institute of Agronomic Research, Regional Center of Agadir, Agadir, Morocco

Kanchan Gandhi Independent Researcher and Visiting Faculty Member, School of Planning and Architecture, Delhi, India

Jamal Hallam National Institute of Agronomic Research, Regional Center of Agadir, Agadir, Morocco

Surendra Singh Jatav Department of Economics, Babasaheb Bhimrao Ambedkar University, Lucknow, India

Youssef Karra National Institute of Agronomic Research, Regional Center of Agadir, Agadir, Morocco

Meriyem Koufan National Institute of Agronomic Research, Regional Center of Agadir, Agadir, Morocco

Rajesh Kumar Department of Environmental Science, Central University of Rajasthan, Ajmer, Rajasthan, India

Hayat Lionboui National Institute of Agronomic Research, Regional Center of Rabat, Rabat, Morocco

Chinmoyee Mallik Department of Rural Studies, West Bengal State University, West Bengal, Kolkata 700126, India

Abhishank Mishra Miranda House, University of Delhi, New Delhi, India

Pooja Rathore Guru Gobind Singh Indraprastha University, New Delhi, India

Ruchi Sachan Miranda House, University of Delhi, New Delhi, India

Benjamin C. Sam Department of Natural and Applied Sciences, TERI School of Advanced Studies, Delhi, India

Lydia Sam School of Geosciences, University of Aberdeen, Aberdeen, UK

Anindita Sarkar Center for Development Research (ZEF), University of Bonn, Bonn, Germany

Subrata S. Satapathy Christ Academy Institute of Law, Bengaluru, India

Raman Sharma Advocacy Officer, Federation of Indian Animal Protection Organisations, Bhathinda, India

Shaktiman Singh School of Geosciences, University of Aberdeen, Aberdeen, UK

Shipra Singh Miranda House, University of Delhi, New Delhi, India

Katyayini Sood United Nations University, Bonn, Germany

Nisha Thankappan Centre for South Asian Studies, School of International Studies, Jawaharlal Nehru University, New Delhi, India;
Climate Change Research Division, Lampero Fos Research Consultants Pvt. Ltd, New Delhi, India

Ahmed Wifaya National Institute of Agronomic Research, Regional Center of Agadir, Agadir, Morocco

Abbreviations

AAA	American Anthropological Association
AAP	Aam Aadmi Party
AAR	Accumulation Area Ratio
ADB	Asian Development Bank
AfD	Alternative für Deutschland
AIDMI	All India Disaster Mitigation Institute
ANI	Asian News International
APL	Above Poverty Line
BAPA	Bangladesh Poribesh Andolon Bangladesh
BDA	Bakkhali Development Authority
BI	Base Inventory
BIFPC	Bangladesh India Friendship Power Company
BISRCI	Bangladesh-India Sundarbans Region Cooperation Initiative
BJP	Bharatiya Janata Party
BOB	Bay of Bengal
BPDB	Bangladesh Power Development Board
BPL	Below Poverty Line
BRT	Bus Rapid Transportation
BRTS	Bus Rapid Transportation systems
CBDM	Community Based Disaster Management
CD	Community Development
CDA	Critical Discourse Analysis
CHARIS	Contribution to High Asia Runoff from Ice and Snow
CITIIS	City Investments to Innovate, Integrate and Sustain scheme
CMIP	Coupled Model Inter-comparison Project
CMP	Comprehensive Mobility Plan
CNG	Compressed Natural Gas
COBOL	Common Business-Oriented Language
COP	Conference of Parties
CWB	Conservation of Water Bodies
DDF	Degree-Day Factor

DEM	Digital Elevation Model
DGPS	Differential Global Positioning System
DRR	Disaster Risk Reduction
ELA	Equilibrium Line Altitude
EM DAT	Emergency Event Database
EPR	End Point Rate
EU	European Union
EV	Electric Vehicle
FAO	Food and Agriculture Organisation
FCC	False Colour Composite
FHH	Female-headed households
FLCN	Fund against Natural Catastrophes
FNC	Fox News Channel
GAMDAM	Glacier Area Mapping for Discharge from the Asian Mountains
GBM	Ganga-Brahmaputra-Meghna Basin
GCCTF	Global Climate Change Task Force
GDP	Gross Domestic Product
GE	Google Earth
GHG	Green House Gas
GLIMS	Global Land Ice Measurements from Space
GLOF	Glacial Lake Outburst Floods
GOI	Government of India
GoWB	Government of West Bengal
HH	Household
HKH	Hindu Kush Himalaya
HRIDAY	Heritage City Development and Augmentation Yojana
IAHS	International Association of Hydrological Sciences
ICIMOD	International Centre for Integrated Mountain Development
ICSSR	Indian Council for Social Science Research
IDPs	Internally Displaced Persons
ILO	International Labour Organisation
IMPRESS	Impactful Policy Research in Social Science
INDC	Intended Nationally Determined Contribution
INGO	Inter-Governmental Organizations
IPCC	Intergovernmental Panel on Climate Change
IPT	Intermediate Public Transport
JNNURM	Jawaharlal Nehru National Urban Renewal Mission
JWG	Joint Working Group
KKC	Kisan Call Centre
KSLCDI	Kailash Sacred Landscape Conservation and Development Initiative
LULC	Land Use and Land Cover
MCA	Municipal Corporation of Amritsar
MDDA	Mussoorie Dehradun Development Authority
MLA	Member of Legislative Assembly
MoHUA	Ministry of Housing and Urban Affairs

MoU	Memorandum of Understanding
MPCE	Monthly Per Capita Income
MW	Mega Watts
NASA	National Aeronautics and Space Administration
NGO	Non-Governmental Organization
NIR	Near InfraRed
NIUA	National Institute of Urban Affairs
NPK	Nitrogen, Phosphorus and Potassium
NRC	National Register of Citizens
NSIDC	National Snow and Ice Data Centre
NSM	Net Shoreline Movement
NTPC	National Thermal Power Corporation
OBC	Other Backward Class
PBMS	Punjab Bus Metro Society
PIDB	Punjab Infrastructure Development Board
PMIDC	Punjab Municipal Infrastructure Development Company
PMTS	Public Mass Transport System
PMU	Program Management Unit
PSPCL	Punjab State Power Corporation Limited
PVTG	Particularly Vulnerable Tribal Group
RAAHI	Rejuvenation of Auto Rickshaws in Amritsar through Holistic Interventions
RCP	Representative Concentration Pathway
RGI	Randolph Glacier Inventory
RMSE	Root Mean Square Error
RS-GIS	Remote Sensing and Geographical Information System
RTA	Regional Transportation Authority
SAD	Shiromani Akali Dal
SC	Scheduled Caste
SDG	Sustainable Development Goals
SDT	Sustainable Development in Tourism
ST	Scheduled Tribe
TCC	True Colour Composite
TEK	Traditional Ecological Knowledge
UAV	Unmanned Aerial Vehicles
UN	United Nations
UNDESA	United Nations Department of Economic and Social Affairs
UNDRR	United Nations Office for Disaster Risk Reduction
UNESCO	United Nations Educational, Social and Cultural Organisation
UNFCCC	United Nations Framework Convention on Climate Change
US	United States
USAID	United States Agency for International Development
USD	US Dollars
UTM	Universal Transverse Mercator
WFH	Work from Home

WGI	World Glacier Inventory
WGMS	World Glacier Monitoring Service
WHO	World Health Organisation
WMO	World Meteorological Organisation

List of Figures

List of Tables

Chapter 1
Contextualizing "Risk", "Uncertainty" and "Maladaptation" in the Context of Climate Change

Anindita Sarkar, Shipra Singh, and Ruchi Sachan

Abstract This introductory chapter of the book is written with two important goals in mind. The first aim is to give a conceptual clarity on the concepts of risks, maladaptation, and uncertainty in the field of climate change and second is to explain the complex interrelationships it has with one another. We begin with the understanding that climate change adaptation is urgently required since the negative impact of climate change is felt worldwide, and evidence-based decision-making and policy options is a priority now. However, adaptation to climate change comes with a complex methodological and implementational challenge because it requires communities and individuals to make decisions with a possibility of very long-term consequences on the basis of existing knowledge that is often incomplete and uncertain. Thus, even with rapid advances in our capability to assess and quantify risk, the potential for adverse consequences remains uncertain. Without complete knowledge, risk analysis becomes inapt. Through various examples, we conclude that economically feasible, culturally accepted, and socially just, climate change adaptation plans can take place by acknowledging the challenge of understanding the nature and degree of uncertainty. The only way forward is to continue in efforts to generate evidence to monitor adaptation plans for minimizing the risk of maladaptation.

Keywords Climate change · Risk · Uncertainty · Maladaptation · SDG · Interconnected risk · Risk of maladaptation

---

A. Sarkar (✉)
Center for Development Research (ZEF), University of Bonn, Bonn, Germany
e-mail: anindita@uni-bonn.de

S. Singh · R. Sachan
Miranda House, University of Delhi, New Delhi, India
e-mail: shipra.singh@mirandahouse.ac.in

R. Sachan
e-mail: ruchi.sachan@mirandahouse.ac.in

A. Sarkar et al. (eds.), *Risk, Uncertainty and Maladaptation to Climate Change*,
Disaster Risk Reduction, https://doi.org/10.1007/978-981-99-9474-8_1

1.1 Introduction

The world is experiencing the impact of climate change with changes in average temperature, variability in rainfall, shifts in the seasons, increasing number and extent of slow onset events, and increasing frequency of extreme weather events. Each of these climate events are associated with increasingly complex risk drivers. Climate change and climate variability are realized as one of the major challenges to human development, as they present an amalgamation of risks and multifaceted risk drivers that negatively impact food and water security, livelihoods based on natural environment, and overall socio-economic development and human welfare. One of the measures to tackle the magnitude of this complex problem is to devise methods for adaptation, simultaneously with efforts for its mitigation. To make climate change agenda inclusive with the developmental goals, United Nations has dedicated one of its seventeen Sustainable Development Goals (SDGs) especially for tacking climate change-related impacts on an urgent basis. SDG 13 envisages achieving sustainable development by adapting and integrating climate change measures to development strategies by "combating climate change and its impacts by taking urgent actions" (United Nations 2015). This shows the consensus among countries to develop policies and practices to adapt to climate change on an urgent basis.

With greater dissemination of scientific knowledge, particularly with the successive publication of the IPCC reports since 1990, climate change has been demarcated as the "looming disaster" that has an "ever-increasing impact" on our reality (Ciążela 2021). The United Nations Office for Disaster Risk Reduction (UNDRR) also advocated addressing climate change as one of the disaster risk drivers (UNISDR 2015). Risks can arise from both potential impacts of climate change as well as human responses to adapt and manage it. However, in the context of climate change adaptation, human responses also have associated risks that happen due to the potential for such responses not achieving their envisioned goals, or worse, risk-blind planning leading to "maladaptation". Data and knowledge limitations and inaccuracies may further reduce the effectiveness or outcomes of such adaptation measures posing critical problems to devise climate policy and amplified challenges in its implementation.

Appropriate climate change adaptation requires data and knowledge to minimize uncertainty. It also needs dynamic disaster risk management strategies that are context specific, culturally accepted, and socially desirable. While climate change can be better mitigated by reducing carbon emissions at the global level, adapting to climate change is particularly at the community level, depending on the severity of the problem, awareness about it, and the means to adapt. It is also important to acknowledge that disasters occur at the local level and impact the communities first and hence the adaptation mechanisms also come from the locals. Thus, the longer the delay in implementation of adaptation efforts, the more difficult and expensive it will become to adapt, especially at a time when the pace of climate change is increasing fast.

There is no doubt that adaptation to climate change is a complex issue. It requires individuals and communities to zero in on matters which can have long-term consequences and that too on the basis of incomplete knowledge and information (uncertainty). In such a situation, it often becomes difficult to assess the synergies and trade-offs. Thus, the adaptation strategies to be successful, they need to appropriately account for associated uncertainties, together within decision-making processes to minimize potential maladaptation. In other words, the strategies should be equipped for "low negative-impact" rather than "zero negative-impact". If communities can better realize their exposure to risks and their probable impacts, they can plan, and initiate decisions based on identified trade-offs. So, firstly, risk reduction and adaptation strategies need to be user-friendly in their application, where knowledge and circumstances are meaningful at the local community level. Secondly, the communities should be aware of their vulnerabilities. Such knowledge will help them to decide on future strategies to curb future risks and take necessary actions at present, to safeguard life and livelihood. This will further help to minimize maladaptation and reduce their risk for climate change-related disasters.

Strengthening the science-policy-practice nexus is the key to fostering effective and timely policy decision. Unfortunately, there is often little political will or financial incentive to invest resources in reducing risk and uncertainty, i.e., ensuring that something does not happen. Hence, decision-makers continue to act under significant uncertainty. It will, hence, be important to continuously strive for investigating such uncertainties and the impact of adaptations to generate new evidence to add to the existing knowledge to inform policy decisions. Our book is a small endeavor to add to this knowledge with new scientific evidence, critical thinking, and empirical research that cuts across disciplines.

1.2 Defining the Concepts

1.2.1 Climate Change

IPCC (2007) defines climate change as "the change in the condition of climate either because of natural variability or as a result of anthropogenic activities identifiable by changes in the mean and/or the variability of its properties which persists for extended period, typically decades or longer". This definition has been carried forward in all subsequent IPCC reports (IPCC 2013, 2014, 2023). "Climate change evidence includes changes in temperature and precipitation patterns, variation in trend of rainfall, rise in temperature and overall increase in frequency of extreme and harsh weather such as droughts" (IPCC 2014). Problems arising due to climate change are multifaceted in the sense that local, regional, and global concerns and impacts vary at various scales. This translates into a complex task to formulate plans and policies addressing everything in a holistic way (Lempert et al. 2004).

Climate change has induced changes in the core functions of the Earth's systems, alarming researchers all across the planet. However, the most debatable question over few decades has been the use of terms climate variability and climate change in spite of the standard definition adopted by IPCC (2007). Some scholars such as Ross Gelbspan, Sharon Begley (Bloomsbury Press 2010) argue that climate change is a more stringent terminology keeping in mind the fact that Earth's climate has been always varying since its evolution. So, a change in climate is normal. However, the other section of researchers is of the view that the changes noticeable over the last three decades are quite dominant and permanent in nature as with all the development activities, these changes seem irreversible, hence climate change is the suitable term (IPCC 2014). IPCC 2021 has reiterated that human activities are majorly to blame for the changing climate.

The ever-changing nature of Land Use and Land Cover (LULC) has led to various environmental, ecosystem, water, food security, climate change problems of varying degrees in different parts of the globe (Turner BL II et al. 2007). One-third of the global land use has seen changes either once or on multiple occasions over the past few decades (Alkama and Cescatti 2016). The range and scale of these changes determine its consequences on the surroundings and its inhabitants. Thus, ever-changing climate and its underlying association with growing use and abuse of resources have become inherently embedded in the system. Irrespective of the level of development in a region, pressure on resources has been constantly increasing leading to imbalances in the system. It has created an unprecedented threat for the Earth's inhabitants and its environment. The impacts are adverse, be it rising temperatures, melting glaciers, sea-level increase, change in weather patterns, increase in the severity of events leading to diminished crop yields and loss of livelihood (UNFCCC 2019). As the evidence for the impacts of climate change becomes more compelling, addressing this issue is of growing importance for the well-being of the planet and future generations (Seneviratne et al. 2021) and it has an enormous impact on the socio-economic aspect of individuals and communities (Lempert et al. 2004).

While mitigation of climate change by reducing CO_2 emission has been an ongoing negotiation between countries, adaptation has become a critical component to cope with negative effects of climate change that impact the well-being of communities from local to global level (Lempert et al. 2004). Communities with existing socio-economic vulnerabilities are bearing the heaviest brunt of climate-induced disasters cascading these vulnerabilities manifesting into multiple crises including conflict, loss of livelihoods, and economic shocks. Adaptation strategies are now being advocated to ensure a sustainable, holistic, and secure future for all.

Climate change has profound, cyclic, and multifaceted impacts on the Sustainable Development Goals (SDGs) established by the United Nations in 2015. Integration across the SDGs is what is aimed at today by all sectors for achieving the end result. These 17 interlinked goals aim to address a wide range of global challenges, from poverty reduction to environmental sustainability (Lim et al. 2018). Climate change, as a multifold crisis, intersects with many of these goals, both directly and indirectly, and can either enable or hinder their attainment (Griggs et al. 2014). Climate change can impact the other SDGs in the following ways:

- No Poverty (Goal 1)—Climate change can exacerbate poverty by affecting livelihoods through extreme weather events, reduced agricultural productivity, and displacement due to rising sea levels or disasters (Brito 2012).
- Zero Hunger (Goal 2)—Climate change can disrupt food production and availability, leading to increased food insecurity, malnutrition, and hunger in vulnerable regions.
- Good Health and Well-being (Goal 3)—Climate change can impact public health through the spread of diseases, heat stress, and the exacerbation of existing health challenges, particularly in low-income areas (Sustainable Development Solutions Network 2015).
- Clean Water and Sanitation (Goal 6)—Climate change can strain water resources, leading to water scarcity and compromising access to affordable and safe drinking water.
- Affordable and Clean Energy (Goal 7)—The transformation to clean energy sources is necessary for mitigating climate change and achieving energy sustainability, in line with this goal.
- Industry, Innovation, and Infrastructure (Goal 9)—Sustainable infrastructure development should incorporate climate resilience and low-carbon technologies to address climate-related challenges.
- Sustainable Cities and Communities (Goal 11)—Changing climate can result in more frequent and extreme urban disasters, emphasizing the importance of resilient urban planning and infrastructure (Ross 2009).
- Responsible Consumption and Production (Goal 12)—Climate change mitigation involves reducing greenhouse gas emissions, promoting resource efficiency, and shifting toward sustainable consumption and production patterns (Ikuho 2012).
- Climate Action (Goal 13)—This goal directly addresses climate change and highlights the need for global cooperation to reduce emissions and enhance climate resilience.
- Life Below Water and Life on Land (Goal 14)—Climate change negatively affects ecosystems, biodiversity, and ocean health, making these goals essential for conservation efforts.
- Life on Land (Goal 15)—Climate change can lead to deforestation and land degradation, impacting terrestrial ecosystems and biodiversity.
- Partnerships for the Goals (Goal 17)—Climate change requires international collaboration and partnerships to mobilize financial resources and technological support for long-term adaptation plans. Global partnership is needed for mitigation climate change by reducing greenhouse gas emission.

Climate change is a crisis multiplier. Climate change can either facilitate or hinder progress toward sustainable development, depending on the actions taken. Addressing climate change is not just a goal in itself but a cross-cutting issue that influences various aspects of human development, requiring integrated approaches to achieve the SDGs (Chichilnisky 1997). To ensure the success of the SDGs, it is essential to recognize the intricate relationship between human choices and actions,

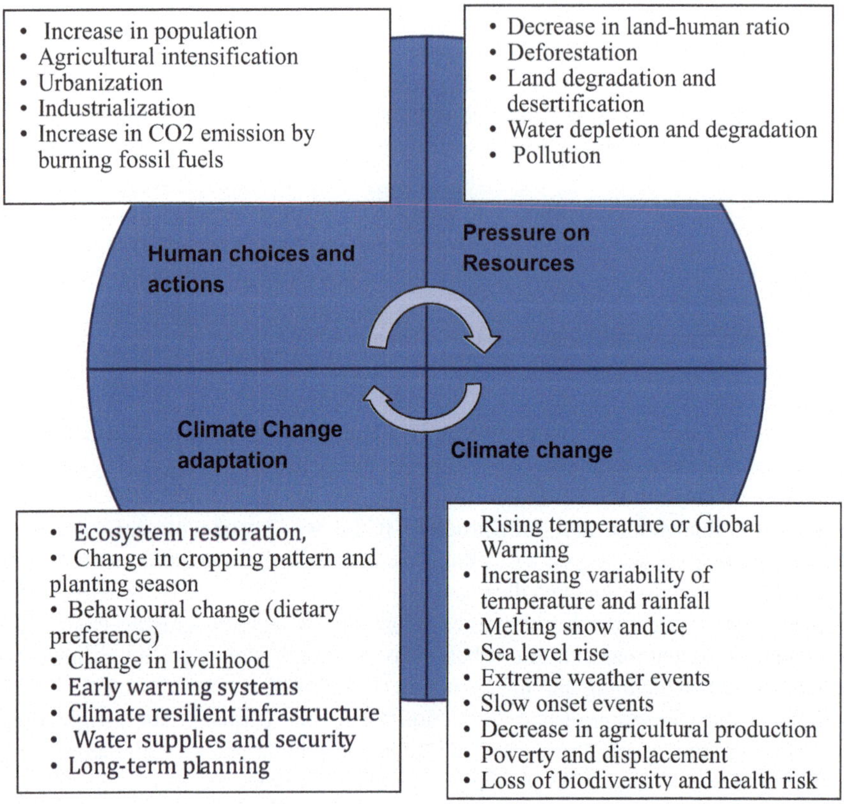

Fig. 1.1 Vicious cycle of human action and choice, climate change and climate change adaptation. *Source* Prepared by Authors

climate change, and climate change adaptation (Fig. 1.1) to take comprehensive measures to build a more sustainable and just future.

The debate on climate change between the Global North and the Global South has been a longstanding and contentious issue within the realm of international environmental policy. The Global North, typically comprising of developed nations, has historically been the largest contributor to greenhouse gas emissions through urbanization, industrialization, and massive energy consumption (Hurrell and Sengupta 2012). In contrast, the Global South, made by coalescence of many developing and less industrialized nations, has contributed significantly less to global emissions but frequently bear the brunt of the most severe consequences of climate change. This ongoing conflict over responsibility and equity in addressing issues of climate change has led to this core and fundamental debate among the two parties (Paterson and Grubb 1992). Global South countries argue that the Global North bears the historical responsibility for the majority of emissions and should take the lead in adapting and mitigating climate change through substantial emission reductions and financial assistance to vulnerable nations (Beer 2014). In contrast, Global North countries emphasize the role of emerging economies in the Global South, such as China and India, as significant contributors to current emissions, adding complexity to the debate.

One crucial aspect of this debate is the disproportionate burden of climate change adaptation that falls on the Global South (Von Bassewitz 2013). Developing nations in the Global South are often more vulnerable and at risk to the impacts of climate change, such as extreme weather events, increase in the frequency of floods and droughts, sea-level rise, and disruptions to agriculture and water resources. These vulnerabilities arise from factors such as larger population, more dependency on primary sectors, and limited resources for disaster preparedness (Haibach and Schneider 2013). As a result, Global South nations face substantial costs and challenges in adapting to the changing climate. The burden of adaptation often requires significant financial resources and technical support, which some argue should be provided by wealthier nations of the Global North as part of their historical responsibility (Never 2013). The global community has recognized the need to address this issue through initiatives like the Green Climate Fund, Clean Development Mechanism Projects designed to help developing nations finance climate adaptation and mitigation efforts. However, ongoing debates persist regarding the adequacy of support and the fair distribution of responsibilities in a world grappling with the consequences of climate change (Ruppel 2013).

While the formation of the loss and damage fund in COP 27 is a significant achievement for hazard-prone developing countries and communities who are most vulnerable to the climate crisis, there seems to be a long road ahead. Establishment of this fund is advocated as the third pillar to the world climate finance landscape, the other two being the mitigation funding that aims to reduce emissions, and adaptation funding that aims to lessen the negative impacts of emissions. The loss and damage funding will address and compensate for the harms caused by emissions. However, several questions remain unanswered, like who will invest money in this fund and in what proportion and how will it be established that a loss or a damage is particularly attributed to climate change. Meng et al. (2023) reported that carbon dioxide (CO_2) emissions have more than doubled from 1995 to 2015 in developing countries. These emissions accounted for 42.8% of global CO_2 emissions in 2015, almost twice the amount emitted by developed countries. The other important fact according to recent data by UNDP (2023), nearly two-thirds of the total poor population (730 million people) is hosted by the middle-income countries and more than 18% live in acute multidimensional poverty across 110 countries of the world. Sub-Saharan Africa and South Asia are home to approximately five out of every six poor people in the world. Eventually, if our global goal is for sustainable development, the climate finance should address the climate crisis taking into consideration this inequality and injustice, not only between countries but also between individuals and communities living in the same country.

1.2.2 Maladaptation

Adaptation broadly refers to the actions that are taken to cope with the negative impacts of climate change. It also includes strategies and policies that are implemented to effectively reduce risk and vulnerability of population exposed to the negative effects of climate change. As Adger et al. (2007) point out that adaptation includes both "anticipatory" and "reactive" actions. It can be a small change in behavior of an individual or a vast infrastructure project funded by the government or change in the cropping pattern of a village. A suitable adaptation practice ideally should be reflected in positive impacts on social, ecological, environmental, and economic life of communities that work to mitigate adverse impacts of climate change and even, if possible, take advantage of arising opportunities from that (Easterling et al. 2000).

Unfortunately, whatever the intention be, the "adaptation" may not be always a success or may become unsuccessful at some point. There are possibilities that it may/ can fail in its basic objective "to reduce risk and vulnerability to climate change" or even make things worse. This is termed as "maladaptation". In other words, maladaptation occurs when adaptation actions or investments made to facilitate an action taken to evade or lessen the impacts of climate change that increases vulnerability to other risks and have an adverse impact on the system or people, or increases the vulnerability of other systems, sectors, or social groups as a collateral damage (Barnett and O'Neill 2010).

The IPCC defined maladaptation in the Third Assessment Report as "any changes in natural or man-made systems that inadvertently increase vulnerability to climatic stimuli; an adaptation that does not lead to reducing vulnerability but increases it instead" (McCarthy et al. 2001: 990). Though the term "maladaptation" is mentioned and referred to in the subsequent assessment reports, it is only in the Fifth Assessment that an entire section is devoted to "Addressing Maladaptation" (Noble et al. 2014). The IPCC Sixth Assessment Report (AR6) supported a more multifaceted view of (mal)adaptation by linking it to both Sustainable Development Goals and climate risks. In other words, it emphasized on a systematic assessment of a set of adaptation responses for climate risk reduction that may affect broader sustainability goals such as equity and justice, to assess synergies, co-benefits, and trade-offs in human and natural ecosystems (Reckien et al. 2023).

Juhola et al. (2016) assert "maladaptation could be attributed to an intentional adaptation policy or measure that can directly increase the vulnerability for the targeted and/or external actor(s), and/or eroding preconditions for sustainable development by indirectly increasing society's vulnerability" (Juhola et al. 2016, 139). This is probably one of the most comprehensive definitions. Even though we do not dwell on all the definitions of (mal)adaptation, it is clear that there is a growing awareness on maladaptation, and that the concept is getting acknowledged rightfully. Or perhaps besides the growing consciousness at both global and local level, the adaptation strategies that were implemented more than a decade has actually started showing "signs of maladaptation".

In the literature two types of maladaptation have been identified. First, the adaptation practices that have "little or no impact" (Anderson 2011; Brooks et al. 2011). Second, some adaptation initiatives that can potentially foster adaptation in the short term, but in the long run they can become harmful negatively affecting territories, sectors, and people's long-term capacity and opportunities to cope with and manage the impacts of climate change impact. These "long-term impacts" are conceptually described as maladaptation (Barnett and O'Neill 2010; Juhola et al. 2016). Here the consensus on "longer term strategy" (Fazey et al. 2016; Wise et al. 2014; Haasnoot et al. 2013) is justifiable since some adaptation strategies have immediate detrimental effect but are better in the long run. However, it is very subjective leaving the duration of timeframe ambiguous.

To justify this gap, several scholars argue that "maladaptation" is not an independent phenomenon, but it is rather a "trajectory of change" (Fazey et al. 2001). In other words, maladaptation has a temporal dimension and hence, maladaptation needs to be viewed from the broader context of "adaptation pathways" (Fazey et al. 2016; Wise et al. 2014; Eriksen et al. 2011) and not just an "independent action". In other words, since adaptation is a process and not a phenomenon, the temporal dimension has a crucial role in evaluating whether an adaptation plan could be detrimental with time (Fazey et al. 2016; Wise et al. 2014; Eriksen et al. 2011). In some cases, uncertainties of local impacts, ecosystems' responses, and society's inability to develop and implement the right options may also lead to maladaptation. Thus additionally, these uncertainties make it further difficult to assess the future effectiveness of adaptation initiatives, as several benefits and adverse effects only appear over a longer duration of time. In this context, monitoring of impacts is of utmost importance to steer a plan in position direction and also to know when to modify and stop.

The effects of an adaptation strategy are based on how the strategy is planned and implemented. In one of the most recent studies, Reckien et al. (2023) found that adaptation options are "rarely fully adaptive or maladaptive", but rather "move along a continuum". This study examined the various types of adaptation strategies undertaken in different parts of the world and concluded that the highest potential for successful adaptation is found in social and behavioral changes of individuals, like change in diets, minimizing wastage of food, and increasing social safety nets. The other successful adaptation actions revolved around ecosystem-based options and nature-based solutions like green roofs, rain gardens, or constructed wetlands. The evidence showed the risks of maladaptation were highest for large infrastructure projects, like the coastal embankments and dams for water storage (Reckien et al. 2023). Even the study by Mallik and Bandopadhyay published in this book recorded similar concerns with respect to embankments in the Sundarbans Deltas to avoid seal level rise and tidal floods.

However, even if (mal)adaptation is not an "incident" but a "continuous process", this leaves a fuzzy line between adaptation and maladaptation that can depend on the time, scales, and local context of the adaptation action, stage of the response process and even on the interpretation. Thus, the IPCC's definition of adaptation as an "adjustment process" (McCarthy et al. 2001) and Barnett and O'Neill S's (2010) view regarding the need to minimize the risks of maladaptation come to light. Thus, several guiding principles are being proposed to prevent maladaptation ranging from decreasing in-situ or ex-situ vulnerability and greenhouse gas (GHG) emissions, reducing detrimental outcomes on justice and equity, and avoiding long-term negative consequences of path dependency effects (Barnett and O'Neill 2010; Magnan et al. 2016; Hallegatte 2009; Magnan 2014).

Maladaptation thus refers to actions, strategies, or policies that are intended to address climate change but result in negative or unintended consequences, often exacerbating vulnerabilities. However, the main concern remains that even now not enough is understood about the nature of initiatives that can be taken to enhance the capability of the socio-ecological systems to adapt in the long run. Though much work on theoretical understandings of maladaptation exists, the major task to move forward remains to start preparing practical guidelines to address clear policy frameworks. The guideline can give a direction to stakeholders and make their tasks easier. Additionally, it will encourage communities to acknowledge that such a risk exists and hence, there is a need to anticipate it even though future adverse effects can neither be anticipated nor avoided completely due to uncertainty. This becomes more complicated as "what is right for one location and context may be completely incorrect for another".

1.2.3 Risk

Risk, in general, means a situation involving exposure to danger. It reflects the probable "negative" effects and unwanted outcome of being susceptible to harm, leading to potential loss (Brown and Damery 2009). But the degree and nature of the loss is not clear and so, the consequences of the activities and their associated uncertainties are two important components associated with the concept of risk (Aven 2019). Thus, risk determines the possibility of effect and exposure (Schnarr and Mertz 2022). Risk in the context of climate change refers to the probability and magnitude of adverse impacts resulting from climate-related hazards or events. It includes the detrimental consequences of climate change on lives and livelihoods, including socio-economic, infrastructural, and environmental assets (IPCC 2020; Kunreuther et al. 2014). IPCC defines risk by "the likelihood of occurrence of hazardous events, with the scale of impact these events are likely to have" (Aven 2019). IPCC also states that "risk equals the potential for consequences where something of value is at stake and where the outcome is uncertain, recognizing the diversity of values" (IPCC 2014). Thus, the risk to climate change impacts is a function of hazard, vulnerability, exposure, and uncertainty.

It is necessary to understand climate risk as the scale of its impact keeps rising over time (Parry et al. 2001). The combination of hazard and vulnerability[1] creates and even increases risk as evident in the "rise of climate-related hazards due to climate

[1] Hazard is an event with the possibility to cause harm and vulnerability refers to the category of being exposed to that possibility, both physically or emotionally (Jones and Boer 2004).

change" (Ireland 2010). Additionally, exposure of any group or region to potential impact of climate change also enhances the degree of risk (Doll and Romero-Lankao 2017). Since vulnerability is governed by sensitivity and adaptive capacity both, the poverty levels and economic competency of any physical space or social group increase the level of exposure to the adverse climate impacts as it lowers the regions or groups adaptive capacity and increases its sensitivity (Woetzel et al. 2020; Doll and Romero-Lankao 2017).

Risks that are being accentuated by climate change encompass the territorial risk of suffering loss, both in terms of lives, livelihood, and property. The "inherent vulnerability" and "nurtured adaptation" both counterbalance each other and contribute to the enhancement or reduction of risk (Briguglio 2010). Two types of climate risks are discussed: "*physical risk*" and "*transition risk*", based on biophysical and socio-economic impact thresholds respectively. The physical risk encompasses the physical impacts, like the occurrence of floods and droughts, prevalence of heatwaves and rising sea level. The transition risk specifies the changes in market, policy, position, and legal status pertaining to alteration in the economic condition (Aven 2019; Jones and Boer 2004; IPCC 2020). With the increase in the frequency and intensity of disasters, transition risk is gaining an enhanced relevance with the increasing levels of socio-economic vulnerability to climate change (Ireland 2010; Woetzel et al. 2020; Aven 2019).

The manner in which climate risk is defined, understood, and used has evolved over time. It now has different theoretical approaches. The two most common approaches are the natural hazard-based approach and the vulnerability-based approach. While the first emphasis on the "biophysical" aspects of climate-related risk, the latter focuses on the "socio-economic" facets (Jones and Boer 2004). In addition to these, there is another stream of studies that has the element of subjectivity attached to it that studies people's perception and awareness (Kunreuther et al. 2014) with respect to climate change and its impact. This is called the psychometric approach (Douglas 1986) that reflects upon the "social construction of risk and its multidimensional complexity" (Etkin and Ho 2007). It engages with ideological, social, and environmental values of people and stakeholders. It also helps us to understand the systematic biases and devise an effective risk assessment strategy to mitigate the adverse impact of climate change.

The importance of risks and uncertainties in determining climate change adaptation lies in their contribution to enhance "socio-natural complexity" and further aggravating it in different regions, both locally and globally (Orderud and Naustdalslid 2018; Hurlbert and Gupta 2016). The uncertainty of the risk also arises from the temporal and spatial changes in the enormity and probability of incidence of hazard, exposure, and vulnerability alike. In such a situation, risk "applies to both impacts of and responses to climate change" (IPCC 2020). While the former relates to the "severity" of impacts that limit adaptation options, the latter pertains to the decision-making in tackling the impact encountered.

1.2.4 Uncertainty

Climate change is a complex, intertwined issue that is fraught with uncertainty. Uncertainty encompasses the lack of complete knowledge or predictability about future climate conditions and their effects (IPCC 2013). It arises from factors like incomplete data, model limitations, and the complexity of climate systems. In the context of climate change, uncertainty is called as a "complex" and a "wicked" problem that makes it hard to initiate and implement policies and adaptation (Incropera 2016). With divergent and incomplete knowledge bases, often problems are construed or framed in very different ways leading to conflict between short-term interests with long-term benefits. In other words, uncertainty can become a root cause of maladaptation. Uncertainty can also become a vicious cycle leading to inaction as without proper proof, adaptation and mitigation measures cannot be framed.

The projections regarding climate change study are heavily based on computer models which require processing of data. These models incorporate various processes such as physical, chemical, and biological to simulate future climate scenarios. Thus, data lags form a major cause of concern in calibrating and validating these models for real-life usage and prediction. Thus, uncertainties arise from incomplete knowledge and evidence about these processes, as well as limitations in data and computing power (IPCC 2013). So, uncertainty can become an aggravating threat that can become disastrous for the whole planet as people, governments, and policy makers tend to undermine the threat due to a lack of data and evidence to support that.

Hawkins and Sutton (2009) put forth in their study that uncertainties in climate study can be caused because of dominance of natural factors overshadowing the actual culprits such as anthropogenic activities. Climate change as defined by IPCC (2007) count changes caused by natural factors such as volcanic activity, lava flow, etc. These factors can divert attention from human-induced changes caused by large-scale deforestation, soil erosion, encroachment of riverbanks, decline in the quality of various ecosystems, etc.

Uncertainty in the system also arises because of the variable and dynamic nature of factors which are counted in for climate modeling. Along with the increase in population and the advent of *technocene*, variations are usually noticed in the level of emissions. These differences are also because of differences in culture and livelihood ways. Hallegatte et al. (2011) outlined the climate risk in developing countries keeping in account the development patterns and changing climate.

For mankind and communities to be aware about the severity of the issue regarding the climate we are breathing in, the first and foremost concern should be to focus on eliminating or reducing the gap in information and knowledge that leads to this uncertainty (Walker et al. 2004). Models which are proven to give rational and specific results should be invested in. Data parity needs to be maintained. Despite the challenges, it is imperative to address climate change head-on through mitigation and adaptation strategies. The prediction regarding climate data for the coming century should be enhanced. While uncertainty is inherent in climate science, it should not serve as a barrier to action. Instead, it should be viewed as a call for careful risk management, adaptive policy-making, and ongoing improvements in our understanding of the climate system. Flexibility in formulating policies should be the norm as the climate has varied and it will continue to do so. Uncertainty has to be accounted in, while planning for a more conscious and sustainable world. In a world where the consequences of inaction are increasingly evident, addressing climate change uncertainty is not just a scientific or policy challenge, but a moral and ethical imperative.

1.3 Interconnections Between "Risk", "Maladaptation", and "Uncertainty"

As climate change adaptation practices have increased across all parts of the world, it is becoming progressively imperative to understand what is working best and what is not working at all. It is because, from the evidence gathered from primary studies from all parts of the world (for eg. see Mallik and Bandopadhyay in this book) it seems many initiatives that are taken in the name of adaptation not only waste financial resources but could aggravate vulnerability, which the action was meant to reduce (Magnan et al. 2016; Noble et al. 2014). This raises the crucial question, "what are the risks of maladaptation?". Working Group II in the Fifth Assessment Report of the IPCC for the first time linked "Risk" to "maladaptation" by stating "actions that may lead to increased risk of adverse climate related outcomes, increased vulnerability to climate change, or diminished welfare, now or in the future" (Field et al. 2014).

It seems that at present maladaptation has been noted as a major risk for the actions taken to reduce the negative effects of climate change (Adger et al. 2005; Barnett and O'Neill 2010; FAO 2010). Besides environmental factors, the risks can also arise from the socio-economic and cultural factors leading to disasters. This is also theoretically and conceptually accepted as disasters are phenomena that are produced by our development choices that are not natural. These choices also determine the risks to disasters as the vulnerability and exposure of ecosystems and societies depend on it. Additionally, risk perception varies with the public's familiarity with the risk (Slovic 2000). Consequently, individual and community knowledge, skills, expertise, attitudes, and the consequent risk behavior are among the key factors that influence vulnerability to disasters (Satya 2010). All these factors impact the behavior and the degree of alarm associated with a disaster or the risk of a disaster. Risk perception interacts with fundamental values to form subjective and mutable limits to adaptation that may either deter or enhance society's ability to act. In other words, psychological, social, and cultural processes can amplify or diminish both community (group) and individual perception of risk and shape risk behavior (Renn et al. 1992). Moreover, collective action in most cases is responsible for influencing individual action in climate change risk management (David and Elise 2007). Thus, a comprehensive understanding of the social context is required to make risk management decisions. This could in turn proscribe adaptation at societal scales (Adger et al. 2009).

Despite growing efforts to adapt to climate change and evidence of maladaptation from various parts of the world, there appears to be limited information about the risk of maladaptation. It is probably because risk reduction is difficult to measure (Dilling et al. 2019; Owen 2020; Singh et al. 2021). Scholars have pointed out the main limitation for risk assessment as the difficulty in recognizing relevant assessment criteria. This is because no specific set of indicators can comprehensively capture the dynamic nature of the risk that depends on the temporal dimensions of responses (long time horizons of outcomes) as well as the context-specific nature of risk drivers and adaptation outcomes (Magnan et al. 2020; Orlove 2022). In other words, risk measurement varies or may vary due to a multitude factors like contradictory adaptation definitions and goals, overlapping of adaptation with development interventions, local values on "tolerable" risk that are specific to a particular geographical place or/and socio-economic context (Singh et al. 2021; Dilling et al. 2019; Owen 2020; Magnan et al. 2020; New et al. 2022; Orlove 2022).

While mitigation efforts may reduce the pace of climate change at the global scale, the primary benefits of adaptation will be generally local. Unfortunately, adaptation to climate risks that involve technical measures and modifications (Wall et al. 2004) may create benefits for some sectors or places, while it may become detrimental to others. In the same context, the resilience scholars debate that building resilience to a specific hazard may come at the cost of increasing the vulnerability to other hazards (Robards et al. 2011). This concept is also being debated by the disaster risk reduction debates that talk of interconnected risks to disasters because "disasters are not isolated incidents but are byproducts of our global systems, existing vulnerabilities, interconnectivity and inequalities that can lead to catastrophes" (UNU EHS 2022 https://interconnectedrisks.org/). This could be applied to maladaptation also because maladaptation is caused when an action to address one specific dimension of risk leads to a new driver of vulnerability. In other words, maladaptation is identified, when evidence of an increase of vulnerability emerges in an action meant to produce positive outcomes. Vulnerability also increases risk and maladaptation to climate change. In other words, maladaptation is a mechanism that negatively affects natural ecosystems and communities by increasing their exposure and/or sensitivity to climate-related stressors. Thus, it requires a multifaceted approach to "effective risk-reduction solutions". Evidently, both nexus approach and interconnectivity lens are needed for building the long-term climate change adaptation plans enabling them to prepare, respond, and cope with climate change disasters.

Ecosystem sensitivity adds another dimension to risk. It is because ecosystems are not in a fixed state and climate change impacts ecosystems at a faster rate, threatening their capacity and resilience with broader environmental changes. Society is also simultaneously affected and can equally affect local environmental dynamics influenced by its decisions and behaviors that lead to vulnerabilities. This makes "exposure" (to risk) dynamic. Since there are no "natural disasters" as discussed earlier, success of an adaptation strategy depends on appropriate decisions. That is why several climate change responses and adaptations now consider both benefits and risks involved. This largely depends on the nature and location of the affected system, magnitude of change, existing vulnerabilities, and values and objectives of different actors.

But is it possible to take an "appropriate decision" based on incomplete knowledge? Moreover, who has seen the future! It is not uncanny that many of the scientific projections prove to be wrong because very often such models are based on incomplete data and knowledge (See Sam et al. in this book). Climate change introduces various forms of uncertainty, including uncertainty in future emissions, temperature changes, and the regional impacts of climate change. Thus, uncertainty can lead to poor decision-making, increasing the risk to maladaptation. In the face of uncertainty, there is a risk of implementing maladaptive measures that do not adequately address the actual impacts of climate change or inadvertently worsen vulnerabilities. In addition to the neglect of the future impacts of climate change and its related uncertainty, often other factors like socio-economic characteristics, cultural values, governance systems are disregard as the main drivers of the system's vulnerability that determine maladaptation. It might cause policymakers to underestimate the risk of climate change or make choices that are overly conservative, costly, or ineffective.

In summary, understanding the connections between risk, uncertainty, and maladaptation in the context of climate change (Fig. 1.2) is crucial for developing effective strategies to mitigate and adapt to climate impacts. It requires acknowledging the inherent uncertainty in climate science, using risk assessment and management tools, and promoting adaptive and informed decision-making processes. Three important interconnections can be inferred between *risk, uncertainty*, and *maladaptation*. First, risk assessment and management processes play a crucial role in addressing uncertainty and avoiding maladaptation. These processes involve identifying risks, evaluating their likelihood and consequences, and developing adaptive strategies. Second, (mal)adaptation is context specific, it is important to consider different scenarios and levels of uncertainty in risk assessments to make informed decisions. Third, adaptive strategies need to be flexible so that they can withstand the consequences of uncertainty and "re-adapt" and not "mal-adapt". In this context, adaptive strategies need to be dynamic that should have enough provision to make adjustments when new information and knowledge comes to light to deal with uncertainties. Unfortunately, some degree of uncertainty will always remain. To factor in that uncertainty, the policy and decision-making will have to underline the value of "low-regret" from "no-regret" strategies that provide benefits even under different climate change scenarios, reducing the risk of maladaptation.

1.4 Importance of Research and Innovation

Research and innovation have two important roles in addressing these challenges. It can help in generating knowledge and methods that can support in taking informed decisions. First, research and innovation in climate science and adaptation strategies can reduce uncertainty and enhance our ability to manage risks effectively. Second,

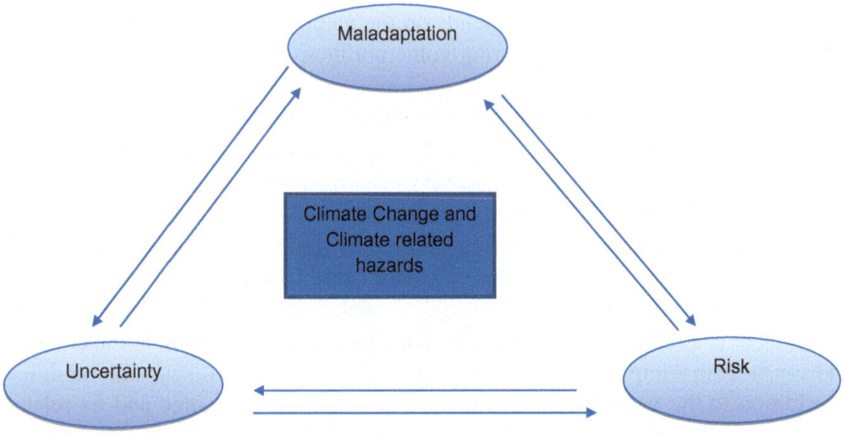

Fig. 1.2 Interconnections between "Risk", "Maladaptation" and "Uncertainty". *Source* Prepared by Authors

research can generate meaningful data for continuous monitoring and evaluation of climate adaptation measures to detect early signs of maladaptation and make necessary adjustments. However, it is important to advocate for transparent communication of uncertainties and potential risks to the public and policymakers to ensure informed decision-making. The scientific knowledge generated when shared with policymakers, researchers, and stakeholders can encourage robust decision-making processes that acknowledge uncertainty and aim to avoid maladaptation.

There are multiple drivers, processes, and dimensions of maladaptation, with its impact and scale varying through ecosystems, social groups, economic sectors, locations, and lifestyles. There is evidence of these multifaceted complexities. A case in study is Coastal-Climate Resilient Infrastructure Project in South-West Bangladesh, where the benefits declined only to leave behind an adverse impact for the inhabitants and the region (Magnan 2016) showcases how short-sighted adaptation strategies lead to long-term damaging effects. Differential vulnerability also increases the uncertainty of risk. Examples can also be drawn from studies in Bangladesh, where the social and cultural norms influences the scale of exposure to which women are subjected to (Bradshaw and Fordham 2013) and how their livelihood security is reduced due to negative consequences of flood control programs with elimination of income opportunities for them (Sultana 2010). Satapathy's research (in this book) further effectively demonstrates the gender inequity in climate impacts and how the indigenous knowledge of Bonda tribal women in Odisha helps in preserving the natural resources and mitigating the challenges of climate change.

Maladaptation can also be due to detrimental institutional policies like agricultural climate insurance which reduced the overall social capital, knowledge base, and risk mitigation capacity of the farmers (Muller et al. 2017). In another instance, building embankments and dykes to protect against rising sea levels and tidal surges in Sundarban deltas of India without considering ecological impacts and geomorphic processes have led to maladaptation as shown in Mallik and Bandyopadhyay (in this book). Similarly, a localized approach to coastal erosion on the west coast of Cape Town resulted in not just biophysical losses including loss of beach area, but also had socio-economic repercussions in the form of loss of tourism and tourism-related revenue (Magnan 2016). A project of developing Hulhumale as a "safe island" in Maldives with an objective to prevent people from sea-related hazards, instead led to deterioration of reefs in the region with large-scale dredging (Magnan 2016). A well-recognized case in point is also the sea-wall project in Fiji that negatively impacted people's lives and livelihoods making them more vulnerable to climate impacts (Piggott-McKellar et al. 2020; Schipper 2020). Gandhi and Sharma (in this book) gave an example of maladapted public transport infrastructure development program in the Amritsar city of Punjab, India. The new transport system adds more CO_2 emissions to the environment instead of lessening it as was intended by the policy makers.

At its very core, maladaptation highlights "rebounding vulnerability", examples of which are well-documented. Adaptive strategies advocating shifting from agricultural practice to industry by farmers for short-term gain in wage and security, leaving farmers with little or no options for return in case of adversity (Schipper 2020) are examples of it. Agreements putting restrictions on fishing to conserve marine ecosystems have a similar effect (Reckien et al. 2023). Also seen in the regional adaptation policy for Afar in Ethiopia where promotion of non-pastoral livelihoods and investment in irrigation agriculture further exposed people to the adverse impacts of climate change, while also diminishing their economic viability (Magnan 2016). The research by Choudhary (in this book) also exposes comparable vulnerability in the hill state of Uttarakhand in India. More importantly, it shows how adaptive actions in the name of "personal healing" as the choice of second homes and retirement homes also induces undue pressure on the environment.

The behavioral aspect of people and places is important as well. For instance, transition from chicken to duck in poultry farming is only beneficial with modifications in people's dietary habits (Schipper 2020). Sood's study also shows that local communities in the villages in Himachal Pradesh invest on adaptation strategies based heavily on local customs and village priests that have and may ultimately prove ineffective due to unforeseen climate impacts. Maladaptation is also a result of ignoring the local realities as highlighted in a study in northern Ghana where migration as an adaptive strategy brought forth newer challenges for the people (Antwi-Agyei et al. 2018).

Discovery and discussion of such knowledge and evidence can lead to more efficient and effective prevention, preparedness, response, and recovery by making more efficient use of financial and human resources. It also helps in risk-based decision-making. In other words, knowledge and evidence results in balanced judgment. The factual evidence in relation to interests and values also helps in reducing uncertainty in decision-making and helps in managing emerging risks that may not be of immediate public concern but may become an important issue during implementation (Renn 2006). Additionally, mobilizing and influencing risk dispositions and behavior of individuals through education and information may also partially transfer the agency of climate change risk control from the state to the individuals. We must also understand that state funding is effectively coming from people's money that comes in the form of tax. Any state funding from taxpayers' money requires legitimization and approvals from its voters. In this way, risk communication can offer a better alternative to regulatory enforcement which may prove more sustainable in the long run both in terms of funding and implementation. Moreover, such communication can be utilized in the prediction, prevention, and remediation of risks. Implicitly, communication is a cost-effective alternative to regulatory systems in climate change action and sustainable development discourses. Science can and must contribute to solutions and the transformative changes needed in the future.

1.5 Conclusion

Understanding of climate change adaptation and its impact on the different levels of decision-making and policy options is a priority now. With multifaceted risks and dynamic nature of ecosystems, adaptation needs to be continuous stream of activities, actions, and attitudes that addresses all aspects of human life and even the social norms (Nelson et al. 2007). Thus, it is important to frame climate change adaptation policies and decision from "social system lenses" as well as "interconnection lens". It is also important to keep enough space and flexibility to maintain a wide range of possible choices when dealing with future climate-related changes.

There is a growing recognition that climate change and risk management strategies should take into account, both risk and uncertainty. There is enough evidence to establish that there are side effects of adaptation initiatives. Unfortunately, decision-makers often rely on spontaneous thinking processes rather than undertaking a systematic analysis of options in a calculated manner. We already have a significant amount of evidence of maladaptation to climate change from various parts of the world and in different contexts which keeps increasing over time. It is now imperative that there is a crucial need for in-depth understanding of the roots and forms of maladaptation since we can foresee that with time the number of adaptation activities will not only grow but will be scaled up and out. We believe for successful solutions to such complex issues will require active and sustained engagement by the governments along with other stakeholders including national, regional, multilateral, and international organizations, the public and private sectors, and the civil society.

In the end, we need to recognize that climate change takes place in the wider context of global transformations. Inadvertently, risks and uncertainty of climate change (mal)adaptation will be influenced by other dynamic factors, like the globalization of economy, political unrest and wars, depletion natural resources due to population pressure, and occurrence of non-climate-related natural hazards like volcanoes and earthquakes. In some such instances, scientists, decision-makers, and stakeholders can do nothing to avoid any anticipative measure. However, it is undefended that both the cumulative evidence on climate change and the limited availability of funding for adaptation is recognized and more data and knowledge will be required for pragmatic and affirmative action. The whole issue of climate finance, loss, and damage will be ultimately based on advocacy and negotiations backed by evidence and knowledge. In line with many other works, the chapters in this book affirm that a challenge for future research consists of developing context-specific strategies that will allow funding bodies to make the best decisions to support adaptation.

References

Adger WN, Arnell NW, Tompkins EL (2005) Successful adaptation to climate change across scales. Global Environ Change 15:77–86. https://doi.org/10.1016/j.gloenvcha.2004.12.005

Adger N, Aggarwal P, Agrawala S, Alcamo J, Allali A, Cruz RV et al (2007) Climate change 2007: impacts, adaptation and vulnerability working group II contribution to the intergovernmental panel on climate change fourth assessment report, Geneva, Switzerland. https://www.ipcc.ch. Accessed 4 Aug 2017

Adger WN, Dessai S, Goulden M, Hulme M, Lorenzoni I, Nelson DR et al (2009) Are there social limits to adaptation to climate change? Clim Change 93:335–354. https://doi.org/10.1007/s10584-008-9520-z

Alkama R, Cescatti A (2016) Biophysical climate impacts of recent changes in global forest cover. Science 351:600–604

Anderson S (2011) Assessing the effectiveness of climate adaptation. IIED Opinion Bulletin. October issue. http://pubs.iied.org/pdfs/17104IIED.pdf

Antwi-Agyei GP, Dougill AJ, Stringer LC, Ardey Codjoe SN (2018) Adaptation opportunities and maladaptive outcomes in climate vulnerability hotspots of northern Ghana. Climate Risk Manag 19:83–93

Barnett J, O'Neill S (2010) Maladaptation. Glob Environ Change 20:211–213

Beer CT (2014) Climate justice, the global south, and policy preferences of Kenyan environmental NGOs. The Global South 8(2):84–100. https://doi.org/10.2979/globalsouth.8.2.84

Bloomsbury Press (2010). Merchants of doubt—How a handful of scientists obscured the truth on issues from tobacco smoke to global warming.

Bradshaw S, Fordham M (2013) Women, girls and disasters—a review for DFID. https://gsdrc.org/document-library/women-girls-and-disasters-areview-for-dfid/

Briguglio, Lino Pascal (2010) Defining and assessing the risk of being harmed by climate change. Int J Climate Change Strategies Manag 2(1):23–34

Brito L (2012) Analyzing sustainable development goals. Science 336(6087):1396–1396. http://www.jstor.org/stable/41585049

Brooks N, Anderson S, Ayers J, Burton I, Tellam I (2011) Tracking adaptation and measuring development. IIED Working Papers 1/2011. http://pubs.iied.org/pdfs/10031IIED.pdf

Brown James D, Damery SL (2009) Uncertainty and Risk in A companion to environmental geography. In: Castree N, Demeritt D, Liverman D, Rhoads B (eds) Blackwell Publishing, West Sussex, UK

Chichilnisky G (1997) What is sustainable development? Land Econ 73(4):467–491. https://doi.org/10.2307/3147240

Ciążela H (2021) The approach of the exact sciences and philosophy towards the looming climate change disaster. Ruch Filozoficzny 77(4):41–56

David E, Elise H (2007) Climate change: perceptions and discourses of risk climate change: Perceptions and discourses of risk. J Risk Res 10(5):623–641. https://doi.org/10.1080/13669870701281462

Dilling L et al (2019) Is adaptation success a flawed concept? Nat Clim Change 9:572–574

Döll P, Romero-Lankao P (2017) How to embrace uncertainty in participatory climate change risk management—A roadmap. Earth's Future 5:18–36

Douglas M (1986) Risk acceptability according to the social sciences. Routledge and Keegan Paul, London

Easterling,DR, Meehl GA, Parmesan C, Changnon SA, Karl TR, Mearns LO (2000) Climate extremes: observations, modeling, and impacts. Science 289(5487):2068–2074

Eriksen S, Aldunce P, Bahinipat CS, d'Almeida Martin R, Molefes JI, Nhemachena C, O'Brien K, Olorunfem F, Park J, Sygna L et al (2011) When not every response to climate change is a good one: identifying principles for sustainable adaptation. Clim Dev 3:7–20

Etkin, David & Ho Elise (2007) Climate change: perceptions and discourses of risk. J Risk Res 10(5):623–641. https://doi.org/10.1080/13669870701281462

FAO (2010) Climate-smart agriculture: agriculture: policies, practices and financing for food security, adaptation and mitigation. 'Climate-Smart' agriculture: Policies, practices and financing for food security, Adaptation and mitigation, FAO, Rome, Italy. https://doi.org/10.1111/j.1467-6346.2009.02662.x

Fazey I, Pettorelli N, Kenter J, Wagatora D, Schuett D (2001) Maladaptive trajectories of change in Makira Solomon Islands. Glob Environ Change 21:1275–1289

Fazey I, Wise RM, Lyon C, Câmpeanu C, Moug P, Davies TE (2016) Past and future adaptation pathways. Clim Dev 8:26–44

Field CB, Barros VR, Dokken DJ, Mach KJ, Mastrandrea MD, Bilir TE, Chatterjee M, Ebi KL, Estrada YO, Genova RC et al (2014) Climate change. 2014: impacts, adaptation and vulnerability. Contribution of Working Group II to the Fifth Assessment Report of the Intergovernmental Panel on Climate Change, Cambridge University Press, Cambridge

Griggs D, Smith MS, Rockström J, Öhman MC, Gaffney O, Glaser G, Kanie N, Noble I, Steffen W, Shyamsundar P (2014) An integrated framework for sustainable development goals. Ecol Soc 19(4). http://www.jstor.org/stable/26269703

Haasnoot M, Kwakkel JH, Walker WE, ter Maat J (2013) Dynamic adaptive policy pathways: a method for crafting robust ecisions for a deeply uncertain world. Glob Environ Change 23:485–498

Haibach H, Schneider K (2013) The politics of climate change: review and future challenges. In: Ruppel OC, Roschmann C, Ruppel-Schlichting K (eds) Climate change: international law and global governance: Volume II: policy, diplomacy and governance in a changing environment, 1st ed., pp 357–374. Nomos Verlagsgesellschaft mbH. http://www.jstor.org/stable/j.ctv941vsk.18

Hallegatte S (2009) Strategies to adapt to an uncertain climate change. Glob Environ Change 19:240–247

Hallegatte S et al (2011) Decision making under climate risk in developing countries. Policy Research Working Paper No. 5568. World Bank

Hawkins E, Sutton R (2009) The potential to narrow uncertainty in regional climate predictions. Bull Am Meteor Soc 90(8):1095–1107

Head BW (2014) Evidence, uncertainty, and wicked problems in climate change decision making in Australia. Eviron Plann C Gov Policy 32(4):663–679

Hurlbert M, Gupta J (2016) Adaptive governance, uncertainty, and risk: policy framing and responses to climate change, drought, and flood. Risk Anal 36(2):339–356

Hurrell A, Sengupta S (2012) Emerging powers, North—South relations and global climate politics. International Affairs (Royal Institute of International Affairs 1944-) 88(3):463–484. http://www.jstor.org/stable/23255546

Ikuho M (2012) What are sustainable development goals? Institute for Global Environmental Strategies. http://www.jstor.org/stable/resrep00768

Incropera FP (2016) Climate change: a wicked problem: complexity and uncertainty at the intersection of science, economics, politics, and human behavior. Cambridge University Press

IPCC (2007) Impacts, adaptation and vulnerability. In: Parry ML, Canziani OF, Palutikof JP, van der Linden PJ, Hanson CE (eds) Summary for policymakers: contribution of working group II to the fourth assessment report of the IPCC. Cambridge University Press, Cambridge, UK

IPCC (2013) Climate change 2013: the physical science basis. In: Contribution of working group I to the fifth assessment report of the intergovernmental panel on climate change

IPCC (2014) Climate change: Impacts, adaptation and vulnerability. In: Field CB et al (eds) The fifth assessment report of IPCC. Cambridge University Press, New York

IPCC (2020) The concept of risk in the IPCC sixth assessment report: a summary of cross-working group discussions. Intergovernmental Panel on Climate Change, Geneva, Switzerland

IPCC (2023) Summary for Policymakers. In: Core Writing Team, Lee H, Romero J (eds) Climate change 2023: synthesis report. contribution of working groups I, II and III to the sixth assessment report of the intergovernmental panel on climate change. IPCC, Geneva, Switzerland, pp 1–34. https://doi.org/10.59327/IPCC/AR6-9789291691647.001

Ireland P (2010) Climate change adaptation and disaster risk reduction: contested spaces and emerging opportunities in development theory and practice. Climate Dev 2(4):332–345

Jones R, Boer R (2004) Assessing current climate risks. In: Lim B, Spanger-Siegfried E, Burton I, Malone E, Huq S (eds) Adaptation policy frameworks for climate change: developing strategies, policies and measures. Cambridge University Press, Cambridge, UK, pp 91–117

Juhola S, Glaas E, Linnér B-O, Neset T-S (2016) Redefining maladaptation. Environ Sci Policy 55:135–140

Kunreuther H, Gupta S, Bosetti V, Cooke R, Dutt V, Ha-Duong M, Held H, Llanes-Regueiro J, Patt A, Shittu E, Weber E (2014) Integrated risk and uncertainty assessment of climate change response policies. In: Climate change 2014: mitigation of climate change. In: Edenhofer O, Pichs-Madruga R, Sokona Y, Farahani E, Kadner S, Seyboth K, Adler A, Baum I, Brunner S, Eickemeier P, Kriemann B, Savolainen J, Schlömer S, von Stechow C, Zwickel T, Minx JC (eds) Contribution of working group III to the fifth assessment report of the intergovernmental panel on climate change. Cambridge University Press, Cambridge, United Kingdom and New York, NY, USA

Laybourn L, Throp H, Sherman S (2023) Transformational change is needed. In: 1.5 °C—dead or alive?: the risks to transformational change from reaching and breaching the paris agreement goal. Institute for Public Policy Research (IPPR), pp 8–11. http://www.jstor.org/stable/resrep47419.5

Lempert R, Nakicenovic N, Sarewitz D, Schlesinger M (2004) Characterizing climate-change uncertainties for decision-makers. Kluwer Academic Publishers. Climatic Change 65:1–9

Lim MML, Jørgensen PS, Wyborn CA (2018) Reframing the sustainable development goals to achieve sustainable development in the Anthropocene—a systems approach. Ecol Soc 23(3). https://www.jstor.org/stable/26799145

Magnan A (2014) Avoiding maladaptation to climate change: towards guiding principles. Sapiens. http://journals.openedition.org/sapiens/1680

Magnan AK, Schipper ELF, Burkett M, Bharwani S, Burton I, Eriksen S, Gemenne F, Schaar J, Ziervogel G (2016) Addressing the risk of maladaptation to climate change. Wiley Interdiscip Rev Clim Change 7:646–665. https://doi.org/10.1002/wcc.409

Magnan AK, Schipper ELF, Duvat VKE (2020) Frontiers in climate change adaptation science: advancing guidelines to design adaptation pathways. Curr. Clim. Change Rep. 6:166–177

McCarthy JJ, Canziani OF, Leary NA, Dokken DJ, White KS (eds) (2001) Climate change. Impacts, adaptation, and vulnerability. In: Contribution of working group II to the third assessment report of the intergovernmental panel on climate change. Cambridge University Press, London and New York

Meng B, Liu Y, Gao Y, Li M, Wang Z, Xue J, Wang K (2023) Developing countries' responsibilities for CO_2 emissions in value chains are larger and growing faster than those of developed countries. One Earth 6(2):167–181

Muller B, Johnson L, Kreuer D (2017) Maladaptive outcomes of climate insurance in agriculture. Glob Environ Chang 46:23–33

Nelson DR, Adger WN, Brown K (2007) Adaptation to environmental change: contributions of a resilience framework. Annu Rev Environ Resour 32:395–419

Never B (2013) Power in global climate governance. In: Ruppel OC, Roschmann C, Ruppel-Schlichting K (eds) Climate change: international law and global governance: Volume II:

policy, diplomacy and governance in a changing environment, 1st ed, pp 217–234. Nomos Verlagsgesellschaft mbH. http://www.jstor.org/stable/j.ctv941vsk.12

New M et al (2022). Climate change. In: Pörtner HO et al. (eds) 2022: impacts, adaptation and vulnerability. IPCC, Cambridge University Press, pp 2539–2654. https://doi.org/10.1017/9781009325844.026

Noble I, Huq S, Anokhin Y, Carmin J, Goudou D, Lansigan F, Osman-Elasha B, Villamizar A (2014) Adaptation needs and options. In: Field CB, Barros VR, Dokken DJ, Mach KJ, Mastrandrea MD, Bilir TE (eds) WIREs climate change addressing the risk of maladaptation to climate change © 2016 Wiley Periodicals, Inc. Chatterjee M, Ebi KL, Estrada YO, Genova RC et al (eds) Climate change 2014: impacts, adaptation and vulnerability. In: Contribution of working group II to the fifth assessment report of the intergovernmental panel on climate change. Cambridge University Press, Cambridge, pp 833–868

Orderud GI, Naustdalslid J (2018) The understanding and role of uncertainty and risk in climate change adaptation: local and central authorities in Norway. Int J Sust Dev World 25(7):579–591

Orlove B (2022) The concept of adaptation. Annu Rev Environ Resour 47:535–581

Owen G (2020) What makes climate change adaptation effective? A systematic review of the literature. Glob Environ Change 62:102071

Parry M et al (2001) Millions at risk: defining critical climate change threats and targets. Glob Environ Chang 11:181–183

Paterson M, Grubb M (1992) The international politics of climate change. International Affairs (Royal Institute of International Affairs 1944-) 68(2):293–310. https://doi.org/10.2307/2623216

Piggott-McKellar AE, Nunn PD, McNamara KE, Sekinini ST (2020) Dam(n) seawalls: a case of climate change maladaptation in Fiji. In: Filho WL (ed) Managing climate change adaptation in the pacific region, climate change management. Springer, pp 69

Reckien D, Magnan AK, Singh C, Lukas-Sithole M, Orlove B, Schipper ELF, Coughlan de Perez E (2023) Navigating the continuum between adaptation and maladaptation. Nat Climate Change 1–12

Reisinger A, Howden M, Vera C et al (2020) The concept of risk in the IPCC sixth assessment report: a summary of cross-working group discussions. In: Intergovernmental panel on climate change, Geneva, Switzerland, p 15

Renn O (2006) Risk communication—Consumers between information and irritation. J Risk Res 9(8):833–849. https://doi.org/10.1080/13669870601010938

Renn O, Burns WJ, Kasperson JX, Kasperson RE, Slovic P (1992) The social amplification of risk: theoretical foundations and empirical applications. J Soc Issues 48(4):137–160. https://doi.org/10.1111/j.1540-4560.1992.tb01949.x

Robards MD, Schoon ML, Meek CL, Engle NL (2011) The importance of social drivers in the resilient provision of ecosystem services. Glob Environ Change 21:522–529

Ross A (2009) Modern interpretations of sustainable development. J Law Soc 36(1):32–54. http://www.jstor.org/stable/40206877

Ruppel OC (2013) Intersections of law and cooperative global climate governance—Challenges in the Anthropocene. In: Ruppel OC, Roschmann C, Ruppel-Schlichting K (eds) Climate change: international law and global governance: Volume II: policy, diplomacy and governance in a changing environment, 1st edn, pp 35–100. Nomos Verlagsgesellschaft mbH. http://www.jstor.org/stable/j.ctv941vsk.8

Satya P (2010) Vulnerabilities to climate change in the agriculture sector, adaptation knowledge platform, AIT/UNEP Regional Resource Centre for Asia and the Pacific, Klong Luang

Schipper E. Lisa F (2020) Maladaptation: when adaptation to climate change goes very wrong

Seneviratne SI, Zhang X, Adnan M, Badi W, Dereczynski C, Di Luca A, Ghosh S, Iskandar I, Kossin J, Lewis S, Otto F, Pinto I, Satoh M, Vicente-Serrano SM, M. Wehner, Zhou B (2021) Weather and climate extreme events in a changing climate. In: Masson-Delmotte V, Zhai P, Pirani A, Connors SL, Péan C, Berger S, Caud N, Chen Y, Goldfarb L, Gomis MI, Huang M, Leitzell K, Lonnoy E, Matthews JBR, Maycock TK, Waterfield T, Yelekçi O, Yu R, Zhou B (eds) Climate change 2021: the physical science basis. contribution of working group I to the

sixth assessment report of the intergovernmental panel on climate change. Cambridge University Press, Cambridge, United Kingdom and New York, NY, USA, pp 1513–1766. https://doi.org/10.1017/9781009157896.013
Singh C et al (2021) Interrogating 'effectiveness' in climate change adaptation: 11 guiding principles for adaptation research and practice. Clim Dev. https://doi.org/10.1080/17565529.2021.1964937
Slovic P (2000) Perception of risk. Routledge, London
Sultana F (2010) Living in hazardous waterscapes: gendered vulnerabilities and experiences of floods and disasters. Environ Hazards 9:43–53
Sustainable Development Solutions Network (2015) Chapter 1: getting to know the sustainable development goals. In: Getting started with the sustainable development goals: a guide for stakeholders. Sustainable Development Solutions Network, pp 5–11. http://www.jstor.org/stable/resrep15866.5
Terje A (2019) Climate change risk—what is it and how should it be expressed? J Risk Res 23(11):1387–1404. https://doi.org/10.1080/13669877.2019.1687578
Turner BL II, Lambin EF, Reenberg A (2007) The emergence of land change science for global environmental change and sustainability. Proc Natl Acad Sci USA 104:20666–20671
UNFCCC (2019) Handbook on the adaptation fund. United Nations Framework Convention on Climate Change
UNISDR (United Nations International Strategy for Disaster Reduction) (2015) Sendai framework for disaster risk reduction 2015–2030. UNISDR, Geneva
UNDP (2023) 25 countries halved multidimensional poverty within 15 years, but 1.1 Billion remain poor. Accessed 24 Oct 2023. https://www.undp.org/press-releases/25-countries-halved-multidimensional-poverty-within-15-years-11-billion-remain-poor#:~:text=According%20to%20the%202023%20release,of%20every%20six%20poor%20people
United Nations (2015) Transforming our world: The 2030 agenda for sustainable development; United Nations: New York, NY, USA
UNU-EHS (2022) Interconnected Disaster Risks report, UNU, Bonn. Accessed 4 Oct 2023. https://interconnectedrisks.org/
von Bassewitz N (2013) International climate change policy: where do we stand? In Ruppel OC, Roschmann C, Ruppel-Schlichting K (eds) Climate change: international law and global governance: Volume II: policy, diplomacy and governance in a changing environment, 1st ed. Nomos Verlagsgesellschaft mbH, pp 101–170. http://www.jstor.org/stable/j.ctv941vsk.9
Wall E, Smit B (2005) Climate change adaptation in light of sustainable agriculture. J Sustain Agric 27(1):113–123
Walker B et al (2004) Resilience management in social–ecological systems: a working hypothesis for a participatory approach. Conserv Ecol 6(1):14
Wise R, Fazey I, Stafford-Smith M, Park SE, Eakin HC, Archer van Garderen ERM, Campbell B (2014) Reconceptualising adaptation to climate change as part of pathways of change and response. Glob Environ Change 28:325–336
Woetzel J et al (2020) Climate risk and response: physical hazards and socioeconomic impacts. McKinsey Global Institute

Chapter 2
Comforting Lies: Authoritarianism, Anti-environmentalism and Climate Change Denial

Abhishank Mishra

Abstract Precise, objective, and comprehensive scientific data has consistently been produced by international organisations such as the IPCC, WMO and NASA, based on the rational expectation that signals have similar meanings and that with greater access to information and scientific data, there will be greater convergence in the appraisal of threat from climate change and stimulate stronger and more concerted efforts amongst nations. Despite consistent diplomatic efforts spearheaded by the United Nations Framework Convention on Climate Change (UNFCCC) and the periodic Conference of Parties (COP) dialogues, there is still persistent inaction and pathological deferment on meeting intended national targets to limit global temperatures. The chapter delves into why despite newer scientific evidence of the anthropogenically induced ill-effects on the environment and increased incidence of severe weather occurrences, there still exists denial, deferment and apathy amongst political leaders and people in making accelerated decisive efforts at mitigating these effects.

Keywords Authoritarianism · Anti-environmentalism · Science related populism · Climate change denial

2.1 Introduction

Under the garb of environmental monitoring and reporting, precise, objective and comprehensive scientific data has consistently been generated by Inter-Governmental Organizations (INGO's) and national agencies such as, World Meteorological Organization (WMO), Intergovernmental Panel on Climate Change (IPCC) and National Aeronautics and Space Administration (NASA) based on the rational expectation that signals have similar meanings, and that with greater access to information and scientific data, there will be greater convergence in the appraisal of threat from climate change and stimulate stronger and more concerted efforts amongst nations. Based

A. Mishra (✉)
Miranda House, University of Delhi, New Delhi, India
e-mail: abhishankmishra@gmail.com

A. Sarkar et al. (eds.), *Risk, Uncertainty and Maladaptation to Climate Change*, Disaster Risk Reduction, https://doi.org/10.1007/978-981-99-9474-8_2

on this data and despite consistent diplomatic efforts spearheaded by the United Nations Framework Convention on Climate Change (UNFCCC) and the periodic Conference of Parties (COP) dialogues, there is still persistent inaction and pathological deferment on meeting intended national targets to limit global temperatures from soaring beyond 2°. This presents us with an intriguing question as to why despite newer scientific evidence of the anthropogenically induced effects on the environment and increased incidence of severe weather occurrences, there still exists denial, deferment and apathy amongst political leaders and people in making accelerated decisive efforts at mitigating these effects.

In searching for explanations for this inaction, the chapter through a political psychology[1] approach engages in critical discourse analysis (CDA). CDA analyzes language as a carrier of meaning and interpretation and concentrates on ways in which meaning making engenders power structures and inequity. This methodology will be suitable in understanding how state rhetoric seeped in denial, selective appropriation and sanctification of scientific facts reproduces and reinforces crony capitalism and income disparities. CDA will also enable a deeper understanding of the perpetual apathy and pathological deferment that is pervasive in political leaders and population at large as it proves effective in situating the context of the discourse within the socio-political structures in which actors are rooted (Fairclough et al. 2013; McCarthy 2021). This study will utilize primary scientific data from environment monitoring and reporting agencies such as NASA, United Nations and IPCC and secondary academic literature engaging with the entanglements between climate change, populism, anti-environmentalism, and authoritarianism.

2.2 Unequivocal Scientific Data on Climate Change

Historically, change in the earth's climate has been an irrevocable fact. In the past 8,00,000 years, eight ice age cycles and warmer periods have occurred. Since the conclusion of the last ice age, 11,700 years ago, ushering in modern climate epoch, changes to the climate have occurred based on subtle alterations in the 'earth's orbital trajectory' and the 'magnitude of energy received from the sun' (NASA 2023). According to the IPCC, since 1970s, when structured and methodical scientific evaluations were initiated, the impact of anthropogenic activity on global warming has progressed from being merely considered theoretical to being settled as an undeniable fact (Hegerl et al. 1996; Ramaswamy et al. 2006; Santer et al. 1996, 2003; Westerhold et al. 2020). From mid-1800's, as a consequence of 'human activities' current warming trend has emerged and is continuing at an unprecedented pace

[1] Political Psychology field of study enquires into how group and individual psychological factors have a bearing on decision making, behavior and institutions. This field of study entails an inquiry of political conduct at the individual or group level within a political system. It engages in understanding ways in which political behavior deviates from rational choice predictions. Psychology as an analytical tool helps in extracting critical insight into the dynamics of mass politics and elite decision making (Huddy et al. 2013).

unseen in recent millennia (Hegerl et al. 1996; IPCC 2021; Ramaswamy et al. 2006; Santer et al. 1996; Santer et al. 2003; Westerhold et al. 2020). The present warming is proceeding at a rate that is ten times quicker than the mean warming registered following an ice age (Gaffney and Steffen 2017; IPCC 2021) and unseen in the last 10,000 years (NASA 2023). Anthropogenic greenhouse gas (GHG) releases have been liable for planet-wide temperatures soaring by 1.1° above levels in 1850–1900 (IPCC 2021). The ramifications of rising global temperatures are evident in surging sea levels, melting ice caps, rising instances of cataclysmic weather events and greater risks (IPCC 2021). It is projected that there is a 50% possibility that earth's temperatures will breach the 1.5 °C barrier between 2021 and 2040 (IPCC 2023). This unfortunate trend could be exacerbated if 'high emission pathways' are not curbed.

The implications of climate change are all pervasive with unprecedented warming of the atmosphere, ocean, and land. Predominantly, driven by carbon dioxide emissions in the atmosphere, average surface temperatures have risen by approximately '2° Fahrenheit or 1 °C' since late nineteenth century (NASA 2023; NCEI 2023; UEA 2023). Glaciers are receding in Himalayas, Alaska, Andes, Alps and Africa and ice Sheets are melting in Greenland and Antarctica, with Greenland losing an average of 279 billion tons of every year in the period from 1993 and 2019 and Antarctica losing 148 billion tons every year (Robinson et al. 2014; Velicogna et al. 2020). A majority of the amplified heat has been absorbed by the ocean with the uppermost 100 m of the ocean displaying warming of 0.67° Fahrenheit since 1969 (Levitus et al. 2017; von Schuckmann et al. 2020). Climate change has also led to rising global sea levels with nearly double the rate of rise by about '8 inches or 20 cm in the last century' and at an exacerbated rate per year (Nerem et al. 2018). With '20–30% of all human induced' carbon dioxide emissions in past decades being soaked up by the ocean (NOAA n.d.). Ocean acidification is threatening aquatic plants and animal biodiversity. Based on these accelerated changes in climate, the frequency of extreme weather events has also increased exponentially (NASA 2023). With projections of 'one third global land area' likely to undergo moderate drought by 2100, and increased incidence of floods, it is projected that on account of drought alone, 700 million people will be on the verge of displacement by 2030 (UN 2023). Around 3–3.6 billion people are situated in areas extremely susceptible to climate change (UN 2023).

2.3 Pathological Deferment in Diplomatic Efforts on Climate Change Mitigation

In a bid to constrain and contain global warming, climate emergency and its ill effects, concerted global diplomatic efforts have been spearheaded by the UNFCCC and the periodic COP meetings annually. One of the most consequential meetings, COP21, the Paris agreement was adopted. Vows were made to limit global temperatures from soaring beyond 2°. It also laid out the 'Intended Nationally Determined Contribution'

(INDC) framework for individual nations outlining their own prospective actions to contain climate change. Through the COP framework under the UNFCCC, in subsequent COP meetings, in Marrakesh (2016), Bonn (2017), Katowice (2018), Madrid (2019), Glasgow (2021) and Egypt (2022), there have been concerted efforts at creating greater synergy between governmental, civil society and non-governmental organizations (COP23 Bonn), on effectively implementing (COP22 Marrakesh) and enabling adaptation efforts (COP 24 Katowice) in consonance to the commitments of the Paris agreement. It was ominously asserted during the Madrid COP25 in 2019, that 'climate emergency was getting worse' (UN 2023). In a bid to constrain temperatures soaring to 1.5° beyond pre-industrial levels, based on stipulations of the Paris accord, GHG emissions will need to reach their zenith prior to 2025 and then plummet by 43% by 2030 and reach net zero by 2050 (UN 2023: 2). Based on Intended Nationally Determined Contributions, it is believed that current national contributions would 'not be sufficient to meet the 1.5° target' (UN 2023: 2).

Despite these consistent diplomatic efforts, there is still persistent inaction and deferment on meeting these goals. These attempts and commitment have been characterized as 'hot air and empty promises' (Varshalomidze and Siddiqui 2021). Global warming trends and rising incidence of severe weather events expose the pathological deferment and dereliction of responsibilities by governments. It shows apathy and inaction towards compelling new scientific evidence of the anthropogenically induced effects on climate change. In COP25, major polluting nations resisted taking initiative to enhance efforts to slow down global temperatures from rising despite the final declaration indicating towards the 'urgent need to cut planet heating greenhouse gasses' and bringing them in line with Paris accords requirements (Washington Post 2019). Crucially, despite G20 country commitments being pivotal to prevent extreme global warming, these countries engaged in tactics of deferment by failing to commit to a 2050 goal to reach net-zero carbon emissions (Agarwal 2021). COP26 met a similar fate with pledges on track to not meet Paris accord targets and the United Nations Secretary General António Guterres asserting deferment would mean 'digging out own graves' (CFR 2023).

It is believed by most experts that current pledges are insufficient and 'not ambitious enough' to restrict global temperatures from soaring by 1.5° (CFR 2023). Despite world leaders such as Seychelles President Wavel Ramkalawan asserting at COP26, the compounding economic loss and increasing magnitude of damage associated with climate change, the biggest emitters such as China committed to less than ambitious goals of 'peak carbon emissions by 2030' and carbon neutrality by 2060 (Varshalomidze and Siddiqui 2021). Along with countries such as Russia and Saudi Arabia having targets that are 'untethered' to 'concrete domestic action plans' (Varshalomidze and Siddiqui 2021), efforts at restricting global temperatures from rising by 1.5° seem bleak. There have also been world leaders, such as Australian Prime Minister Scott Morrison, who have insisted on Climate change being 'a non-event' or 'hoax' by asserting that 'curbs on climate change' should not 'come at the cost of people and businesses.'. These efforts show that world leaders lack integrity and leadership while continuing to evade responsibility (Rigg 2023). Due to 'perceived and immediate negative effects' of action on climate change, there also exists substantial inertia and reluctance amongst political leaders to engage in policies that have long term implications (Nemeth and Olivier 2015: 219). The immanent cost associated with this deferment and denial is the possibility of overshooting 1.5° in early 2030's which will lead to irreversible changes in the environment and ecosystem (IPCC 2023: 2).

2.3.1 Science Related Populism, Corporate Denial Machine and Apathetic Civic Participation

Despite increasing scientific evidence indicating towards anthropogenic causes to climate change, there exists apathy, denial, and inaction amongst people to engage in conscientious collective action or decisively rally around climate change mitigation efforts. This indicates that discussing climate change issues through scientific evidence that is rife with 'academic jargon' may not be the most effective strategy (Moser and Dilling 2011; Nemeth and Olivier 2015: 201). With populist politics becoming more pervasive and taking the form of science-related populism (Mede and Schafer 2020), there have been increasing attempts at characterizing experts or those espousing climate change as 'elites' (Meyer 2023) or 'ivory tower' (Nemeth and Olivier 2015: 221), as 'antagonists of the ordinary people' (Mede and Schafer 2020: 481) and attacking them in the 'name of the people' (Meyer 2023) or 'the frontline' (Meyer 2023). Scholars such as Harry Collins and his co-authors have indicated how with 'the rise of populism in the west', there have been onslaughts on 'scientific expertise' as it exists as 'one of the checks and balances' (Collins et al. 2020: 1–3). In populist discourses, environmentalists and experts are deemed as 'unresponsive rootless technocratic international elites' and alleged of being proponents of strategies and policy frameworks that operate 'against the interests of the people' and 'common sense' (Machin and Wagener 2019). These discourses create chasms between the common citizenry and 'holier than thou' experts (Nemeth and Olivier 2015: 222). These tendencies feed into the natural tendency of resistance to authority, people's mistrust of science and top-down leadership (Nemeth and Olivier 2015).

The pervasiveness of science related populism has allowed manipulation of people and inaction through a 'corporate denial machine'. It has emerged as an effective mechanism through which complacency and inaction breeds (Dunlap and Mccright 2011; Nemeth and Olivier 2015). Denial is at work at the individual, societal and organized level (Dunlap and Mccright 2011). There is an elaborate 'climate change denial machine' at play to promote greater inaction and leave people scrambling for credible information to act upon. There are ample instances of science distorters with special interests, filtering and demonizing the truth to benefit from ill-informed public and policy makers. Selective appropriation and sanctification of scientific facts and concepts by distorters for their agenda at the cost of real science and public good is at play (Dunlap and Mccright 2011). An important instance of this is the

Union of Concerned Scientists (2007) stating that ExxonMobil was funding independent fronts and organizations to engage in information laundering desired messages through promoting peer-reviewed scientific findings, breed skepticism of scientific opinion and raising doubts about most indisputable scientific evidence (Dunlap and Mccright 2011). Gabriel Sherman's (2014) work 'the loudest voice in the room' chronicles how news is distorted to muddle science and reinforce ultraconservative views (Sherman 2014). An archetype of this is the Fox News Channel (FNC) in America, which promotes right wing conservative views, works towards distorting the truth, polarizing audiences and engages in biased reporting (Sherman 2014). In terms of climate change, this channel has been accused of projecting climate change as a 'worldwide conspiracy' by foreign powers to take over American resources and discrediting scientific opinion openly. Systematic deception and misinformation has also been at play with institutions such as the Heartland Institute, a conservative public policy think tank backed by Exxon Mobil and Philip Morris in Chicago concentrating its efforts to discredit the science of anthropogenic causes of climate change, fund climate skeptics and promote climate change skepticism (Gillis 2012). According to Hamilton (2010), denial is pervasive not due to a deficit of information but an excess of culture (Gowdy et al. 2011; Hamilton 2010). In order for politics to prevail, an illusion of safety has been created through denial with a 'misplaced faith in technology and willful forgetting' (Kelman 2007; Nemeth and Olivier 2015). Through smart devices and fake news, a false sense of safety is being perpetuated, modern societies have transformed themselves into 'risk denying and risk averting' cultures that 'have the same effect -to increase the cost of risk to the point where risks really do become irrational' (Scruton 2012).

It was believed by scientists that providing greater access to information would lead to behavioral change (Garrison Institute 2012). However, a natural consequence of greater magnitude of information has been driving people towards culturally determined views, the development of a strong confirmation bias and motivated reasoning (Nemeth and Olivier 2015). Distressing information is also pushing people towards 'information avoidance' (Howell et al. 2014). There is ineffective communication with respect to climate change scenarios as 'gloom and doom' scenarios interfere with individual comfort levels and cause distress leading to people completely tuning out. Progressivist movements and environmental regulations that are threatening change in the way of life and established position of a lot of privileged communities such as White race in America is breeding a preference for staying in the nostalgia of the past and finding security in the status quo (Daggett 2018). Against this ignorance and apathy, a uni-dimensional 'account of scientific expertise and linear model of scientific advice' premised on a nostalgia of the past where there was greater confidence and consensus on truth (Meyer 2023) has been highly ineffectual.

2.3.2 *Rising Tendencies of Authoritarianism, Populism and Anti-environmentalism*

Not coincidentally, with the rising frequency of catastrophic weather events due to environmental degradation, there has been a trend towards rising global authoritarianism and anti-democratic governance. This new variant of authoritarianism that has developed out of late capitalism and neoliberal values is a 'free market authoritarianism' which engages in anti-environmentalism due to dirty entanglements between politicians and big businesses. Authoritarian leaders in catering to corporate interests and consolidating their power have engaged in anti-democratic practices such as drying up allocation to public education and health, censorship of free speech and press, harsh use of law enforcement agencies, undermining democratic institutions and judicial independence (Darian-Smith 2022). More importantly, in the realm of environmental regulation to serve corporate interest, deregulation and privatization or 'corporate takeover of regulatory agencies' has happened (Telesca 2020). Similar to Turkey and Hungary, there is also a trend towards centralization of power with strong men and charismatic leaders turning democratic setups into untamed power of majority (Krastev 2021). Another catalyst in this change was the COVID-19 pandemic, which accelerated the global shift in the direction of authoritarianism, as under the garb of securing public health, authoritarian leaders consolidated power through declaring state of emergency, heightened surveillance, and diminished political and social opposition (Darian-Smith 2022). Democracy organizations and Nobel laureates asserted that COVID-19 was a smokescreen for amplifying authoritarian governance and pulling apart liberal democratic principles silencing critics and decreasing oversight (NDI 2020).

With the rise of radical right-wing parties in Europe, Latin America and Asia, leaders such as Bolsonaro in Brazil, Scott Morrison in Australia and Trump in United States assumed power. Neoliberal policies and fundamentalist thinking enabled leaders such as Trump (America), Putin (Russia), Erdogan (Turkey), Duterte (Philippines) and Orban (Hungary) to make state power servile to global capitalism (Darian-Smith 2022). Anti-environmental policies have openly been endorsed in these countries with environmental regulation being diluted to appease corporate backers and institutions being systematically dismantled. For instance, Donald Trump during his tenure openly denied climate change, dismantled a substantial number of regulations pertaining to the environment such as the clean air act, clean power plan, and the clean water act, withdrew from the Paris accord and instituted at the helm of the Environmental Protection Agency an astute climate denier in Scott Pruitt (Daggett 2018). Apart from deregulation, Trump appeased corporate backers from coal and oil industries by lifting moratorium on federal property on new coal leases, cutting funding on research on ill effects of mountaintop coal removal and allowing offshore drilling in US coastal waters (Daggett 2018). A collusion between these strong men and corporations has also meant a systematic erosion of democratic institutions and regulations. This can amply be seen with Bolsonaro in Brazil allowing access of amazon forests to agricultural businesses. Morrison in Australia, assuring new open pit mines for coal industry. Trump doling out corporate tax cuts to businesses and loosening environmental regulations and enabling access to national parks and public lands for mining and drilling purposes. There has been a concerted strategy of commingling ultranationalism, isolationism, and anti-environmentalism, which involves normalizing corporate greed and repudiating harmonious global futures. With authoritarians such as Bolsonaro shirking and denouncing assistance from France for Amazon rainforest fires in 2019 and 2020 by deeming it foreign interference and asserting that it respects its sovereignty (Darian-Smith 2022). Other similar anti-environmental strategies employed by Trump, Bolsonaro, Morison, Duterte was supporting climate skeptics and containing any kind of climate hysteria and activism (Maza 2019; Rosane 2019; Serhan 2021).

Populist leaders breed scientific skepticism and engage in complete rejection of scientific claims. They argue towards opening up both decision making and truth speaking authority to popular counter-knowledges and counter-expertise that was earlier monopolized by scientific elites (Meyer 2023). Populist leaders in a bid to have unchecked power have questioned the power claims made by scientists through scientific knowledge and data that they have produced. They have tried to question the 'decision making sovereignty' of science and have tried to criticize the 'truth claims' and 'truth speaking sovereignty' (Mede and Schäfer 2020). These leaders have indused mistrust in their countries through asserting that elites in the academic realm exercise illegitimate sovereignty and abuse this power to advance their private interest and ideological agenda rather than forwarding 'objective scientific norms' (Mede and Schäfer 2020: 482). They assert the primacy of 'the people' and science grounded in relevant practical common nonsensical notions (Mede and Schäfer 2020: 483). Populist leaders also question the 'truth speaking sovereignty' of science and attempt to undercut scientific autonomy at truthful knowledge production through asserting aesthetic distance from routine ordinary people's experience (Mede and Schäfer 2020: 483). Populists question scientific expertise both in terms of power claims and truth claims (Meyer 2023). This was demonstrated during Donald Trump's presidency, where he periodically through his scathing rhetoric and alt-facts denied climate change publicly and at the same time through his administration engaged in silencing scientists enlisted by the government, denying both power and the truth claims to the scientific community. In another instance, climate change deniers and sceptics cooked up a conspiracy out of emails from climate scientists at East Anglia University that were leaked and made public with allegations unsupported by proof of falsifying data (Brown 2014). In a bid to strengthen their authority and power, they were alleged to have finessed or air-brushed their public data in this scandal under the apprehension that signs of uncertainty would weaken their climate change advocacy (Brown 2014: 139). Another tactic that right wing conservative parties and populist politicians employed to weaken truth and power claims of science was through pitting popular counter knowledge against objective science. For instance, the "Alternative für Deutschland" (AfD) party in Germany engaged in appealing to alternative climate expertise rather than rejecting climate science. This involved using conspiracy theories as opposing information compared to widely accepted views (Boecher et al. 2022: 834–835; Machin et al. 2017). These tactics were effective due to 'popular suspicion of organized power' in this context scientists working on climate change (Brown 2014). This anti-environmentalism and climate denial has come at significant costs in the global south with farmers, environmental protestors and indigenous people bearing the impact of negative consequences of global climate change (Likhani 2020; Scheidel et al. 2020).

2.4 Follow the Science: Generating Resilience in Diverse Cultural Complexes

In the face of rampant misinformation, rising authoritarianism and populist critiques of objective science, from an anti-populist standpoint, there is a pressing requirement to reassert 'following the science' while the scientific community shows commitment towards opening up decision making authority while retaining truth speaking sovereignty of science (Meyer 2023). There is an emerging need to provide appendages to the people to develop greater insight and elevate their perception. Insight involves understanding 'inner character of underlying truth' (Wolman 1989: 179). There needs to be an attempt at bridging objective facts and subjective values (Garrison Institute 2012: 17).

Despite pro-social reflections and actions not being first level concerns in an individual's hierarchy of needs (Maslow 1943), there definitely needs to be a shift in insight on environmental issues in this regard (Nemeth and Olivier 2015: 219). In order to foster pro-social behavior, grass root communication needs to be reinforced to foster a change in perception and movement towards greater insight. There needs to be effective two-way communication, instead of one-way scientific dictates. There is a strong imperative towards natural scientists collaborating with social scientists to produce necessary change (Nemeth and Olivier 2015: 229). Moreover, in the face climate denial machine, peer-to-peer communication needs to be strengthened. This can be strengthened through stronger local community action, civic engagement, and global activism (Garrison Institute 2013: 4). This also means greater climate literacy and research and development in adaptation technology. Solutions to global warming issues in different geographical locations around the world need to engage in decentralized decision making and participative management, which will require the development of critical thinking capabilities to develop innovative solutions and implementing adaptive solutions (Nemeth and Olivier 2015: 231).

The fact that 'human conduct is an outcome of intricate interaction and interplay of genetic inheritance, social conditioning, and subjective experience; it needs to be factored in for knowledge dissemination and popular perceptions (Nemeth and Olivier 2015: 230). Information needs to be made more palatable to where people are situated 'culturally, emotionally and intellectually' (Garrison Institute 2013). The climate crisis needs to be framed in terms of people's deepest values, beliefs, fears, and experiences along with keeping in mind their core values of self-sufficiency, leadership, progress, financial stability and creating opportunities for employment, asserting the benefits of co-action (Nemeth and Olivier 2015: 222). The strong entanglement between environment and spirituality can also be exploited based on the fact that religious groups have taken up starring roles in most social change movements (Garrison institute 2012: 19). Climate change can only be mitigated by collaborative efforts of both progressive strands in community that are driven by facts and conservatives who endorse cultural values (Nemeth and Olivier 2015: 229).

Disasters such as Hurricane Katrina, deep water horizon oil spill in US gulf coast, Haiti and Philippines earthquakes are testimony to show how it takes years for resolution and recovery from disasters (Nemeth and Olivier 2015: 215). There needs to be development of 'resilience' in the face of climate catastrophes as it builds the ability of a system to withstand disruptions and reorient itself whilst undergoing change. Development of a positive capacity amidst climate shocks also requires withstanding stressors and dealing with trauma (Adger et al. 2011; Walker and Heffner 2010). However, resilience needs to be developed at the level of people, communities, cities, nations, and climates (Nemeth and Olivier 2015). Despite there being an inherent capacity for hope, there are some that lose capacity to generate hope (Lopez et al. 2004; Snyder et al. 2000). Hope needs to be generated amongst people for them to constructively engage with adversity through goals and strategies (Nemeth and Olivier 2015: 214). The most effective strategy to deal with climate distress is seeking solutions and dealing with calamity with resilience (Nemeth et al. 2012). Moreover, complacency and inaction stemming from the threat from climate change being perceived as impersonal and distant an issue needs to be re-casted as a more immediate and personal issue. This self-evidently means giving up common business-oriented language (COBOL) (De Jager 1999). Moreover, people need to be woken up from the illusion of safety created by 'misplaced faith in technology and willful forgetting' (Kelman 2007).

2.5 Conclusion

Despite unequivocal evidence of anthropogenically induced climate change and greater access to information and scientific data, inaction, and deferment stem from the pervasiveness of science related populism which disparages the positionality of experts as elites and antagonistic to the interests of the common people. Science related populism has also been used by authoritarian leaders to systematically undermine checks and balances against their decision-making sovereignty. Authoritarian leaders have not only tried to question the decision-making authority of the scientific community but also tried to discredit the truth claims that these technocratic elites have made to further consolidate their preponderant position. Authoritarian leaders have also toyed with mass psyche by feeding into the hedonistic trait of seeking comfort in times of distress by feeding them comforting lies. This has, however, been at the cost of weakening intended nationally determined contributions, weakening regulations, increasing frequency of catastrophic extreme weather events, and exposing the most vulnerable sections of all countries to disaster.

Inaction and deferment have not only stemmed from myopic political imperatives of leaders but also due to their dirty entanglements with corporations. In a bid to appease their corporate patrons, authoritarians have engaged in systematic anti-environmentalism, which entails deregulation, disinformation, denial, and dismantling institutions and checks and balances. Corporations have played their part in concerted efforts at funding institutions that produce alt-facts, peer-reviewed articles denying climate change and news agencies that spread fake news. In extreme cases people have been pushed towards information avoidance. Based on comfort seeking nature of all individuals, access to overwhelming and distressing objective information has meant people engaging in motivated reasoning, action aligned with their strong confirmation bias and culturally determined views.

In the face of a citizenry being in denial, there is a strong need to build resilience in communities with goal oriented and strategic action. A way back on track is targeting cultural complexes in which people are situated, making climate change concerns more personal and immediate for them. People need to find a way to use digital pathways and creative forms of resistance to hold their leaders accountable, reinforce institutions, and regulations. People through their consumer choices need to sway businesses to make more sustainable, eco-friendly and carbon neutral products. There needs to be a strong initiative back towards 'following science' in a bid to counter climate insecurity that is starting to set in. Privileged subjectivities that no longer feel part of the narrative of progress of the nation and feel alienated due to progressivist tendencies need to be reintegrated in the path toward a common global sustainable future.

References

Adger WN, Brown K, Waters J (2011) Resilience. In: Dryzek JS, Norgaard RB, Schlosberg D (eds) The Oxford handbook of climate change and society New York. Oxford University Press, NY, pp 696–710

Agarwal K (2021) Leaders at COP26 work past deadline—But a climate deal remains out of reach. The Wire, November 13. https://thewire.in/environment/leaders-at-cop26-work-past-deadline-but-a-climate-deal-remains-out-of-reach

Boecher M, Zeigermann U, Berker LE, Jabra D (2022) Climate policy expertise in times of populism–knowledge strategies of the AfD regarding Germany's climate package. Environ Politics 31(5):820–840

Brown MB (2014) Climate science, populism, and the democracy of rejection. In: Crow DA, Boykoff MT (ed) Culture, politics and climate change: how information shapes our common future. Routledge, New York, pp 129–45

Collins H, Evans R, Durant D et al (2020) Experts and the will of the people: society, populism and science. Palgrave Pivot, Cham

Council on Foreign Relations (CFR) (2023) UN climate talks: a timeline. Council on Foreign Relations. https://www.cfr.org/timeline/un-climate-talks

Daggett C (2018) Petro-masculinity: fossil fuels and authoritarian desire. Millennium 47(1):25–44

Darian-Smith E (2022) Global burning: rising anti democracy and the climate crisis. Stanford University Press

De Jager P (1999) Y2K: so many bugs... so little time. Scientific American 280(1):88–93

Dunlap RE, McCright AM (2011) Organized climate change denial. In: The Oxford handbook of climate change and society, vol 1, pp 144–160

Fairclough N, Mulderrig J, Wodak R (2013) Critical discourse analysis. In: Wodak R (ed) Critical discourse analysis: Volume I: concepts. Sage, History, Theory, Los Angeles, pp 79–101

Gaffney O, Steffen W (2017) The Anthropocene Equation. The Anthropocene Review 4(1):53–61

Garrison Institute (2012) Comprehensive report on the 2012 climate, mind and behavior initiative. https://www.garrisoninstitute.org. Accessed 3 Apr 2015

Garrison Institute (2013) 2013 Climate, mind and behavior symposium: synthesis report: The human dimensions of resilient and sustainable cities. https://www.garrisoninstitute.org/climate-and-behavior/transformational-ecology-reports. Accessed 03 April 2015

Gillis J (2012) Clouds' effect on climate change is last bastion for dissenters. The New York Times, April 30. http://www.nytimes.com/2012/05/01/science/earth/clouds-effect-on-climate-change-is-last-bastion-for-dissenters.html?_r=0. Accessed 18 April 2015

Gowdy J, Bell RG, Krall L, Walton M (2011) Garrison Institute Report. The social brain and the diffusion of pro-social behavior: background paper for the Garrison institute climate, mind and behavior program. Garrison Institute, Garrison, NY

Hamilton C (2010) Appeasing climate denial at the ABC: on journalistic balance and the war on science. Overland 200:73–78

Hegerl GC et al (1996) Detecting greenhouse-gas-induced climate change with an optimal fingerprint method. J Clim 9:2281–2306

Howell J, Crosier B, Shepperd J (2014) Does lacking threat-management resources increase information avoidance? A multi-sample, multimethod investigation. J Res Pers 50:102–109

Huddy L, Sears DO, Levy JS (2013) Introduction: theoretical foundations of political psychology. In: Huddy L, Sears DO, Levy JS (eds) The Oxford handbook of political psychology, 2nd edn. Online edn, Oxford Academic. https://doi.org/10.1093/oxfordhb/9780199760107.013.0001

Intergovernmental Panel on Climate Change (IPCC) (2021) IPCC sixth assessment report. Working Group I. Technical Summary. IPCC. https://www.ipcc.ch/report/ar6/wg1/downloads/report/IPCC_AR6_WGI_TS.pdf

Intergovernmental Panel on Climate Change (IPCC) (2023) IPCC sixth assessment report, working group I, Chapter 2: physical climate system and the global energy budget. Cambridge University Press. https://www.ipcc.ch/report/ar6/wg1/downloads/report/IPCC_AR6_WGI_Chapter02.pdf

Kelman A (2007) Boundary issues: clarifying new Orleans's Murky Edges. J Am Hist 94(3):695–703

Krastev I (2021) Biden can't decide what counts as a 'Democracy'. The New York Times, May 12. https://www.nytimes.com/2021/05/12/opinion/biden-democracy-alliance.html

Levitus S et al (2017) NCEI ocean heat content, temperature anomalies, salinity anomalies, thermosteric sea level anomalies, halosteric sea level anomalies, and total steric sea level anomalies from 1955 to present calculated from in situ oceanographic subsurface profile data (NCEI Accession 0164586), Version 4.4. NOAA National Centers for Environmental Information. https://www.ncei.noaa.gov/products/climate-data-records/global-ocean-heat-content

Likhani N (2020) Who killed Berta Caceres? Dams, death squads, and an indigenous defender's battle for the planet. Verso, London

Lopez SJ, Snyder CR et al (2004) Strategies for accentuating hope. In: Positive psychology in practice, pp 388–404

Machin A, Ruser A, Von Andrian Werburg N (2017) The climate of post-truth populism. Public Seminar. http://www.publicseminar.org/2017/06/the-climate-of-post-truth-populism

Machin A, Wagener O (2019) The nature of green populism? Green European Journal. https://www.greeneuropeanjournal.eu/the-nature-of-green-populism/. Accessed 2 Sept 2019

Maslow AH (1943) A theory of human motivation. Psychol Rev 50(4):370–396

Maza C (2019) Far-right climate denial is growing in Europe, New Republic, Nov 11. https://newrepublic.com/article/155669/far-right-climate-denial-growing-europe

McCarthy DR (2021) Imagining the security of innovation: technological innovation, national security, and the American way of life. Critical Stud Secur 196–211. https://doi.org/10.1080/21624887.2021.1934640
Mede NG, Schäfer MS (2020) Science-related populism: conceptualizing populist demands toward science. Public Underst Sci 29(5):473–491
Meyer JM (2023) Power and truth in science-related populism: rethinking the role of knowledge and expertise in climate politics. Polit Stud 144–165
Moser SC, Dilling L (2011) Communicating climate change: closing the science action gap. In: Dryzek JS, Norgaard RB, Schlosberg D (eds) The Oxford handbook of climate change and society. Oxford University Press, New York, NY, pp 161–174
NASA (2023) Evidence: how do we know climate change is real? NASA, Washington D.C. https://climate.nasa.gov/evidence/
NASA Goddard Institute for Space Studies (GISS) (2023) GISS Surface Temperature Analysis (GISTEMP) NASA. http://data.giss.nasa.gov/gistemp
National Democratic Institute (2020) A Call to Defend Democracy, NDI. https://www.ndi.org/publications/call-defend-democracy
National Oceanic and Atmospheric Administration (NOAA) (n.d.) Global Monitoring Laboratory—Carbon Cycle Greenhouse Gases—Trends in Atmospheric Carbon Dioxide. NOAA. https://gml.noaa.gov/ccgg/trends/
National Centers for Environmental Information (2023) Monitoring. National Centers for Environmental Information. NCEI. https://www.ncei.noaa.gov/monitoring
Nemeth DG, Kuriansky J et al (2012) Addressing anniversary reactions of trauma through group process: the Hurricane Katrina anniversary wellness workshops. Int J Group Psychother 62(1):129–142
Nemeth DG, Olivier TW (2015) Perceptions of the environment. In: Ecopsychology: Advances from the intersection of psychology and environmental protection, vol 1, pp 193–217
Nerem RS, Beckley BD et al (2018) Climate-change–driven accelerated sea-level rise detected in the altimeter era. Proc Natl Acad Sci 115(9):2022–2025
NOAA Pacific Marine Environmental Laboratory (n.d.) What is ocean acidification? NOAA. http://www.pmel.noaa.gov/co2/story/What+is+Ocean+Acidification%3F
Ramaswamy V et al (2006) Anthropogenic and natural influences in the evolution of lower stratospheric cooling. Science 311:1138–1141
Rigg S (2023) Two global climate conferences, two failures of political leadership. Action Aid. https://www.actionaid.org.uk/blog/2023/06/26/climate-conferences-failures#:~:text=The%20years%20of%20unfulfilled%20promises,was%20palpable%20at%20the%20conference
Robinson DA, Hall DK, Mote TL (2014) Measures northern hemisphere terrestrial snow cover extent daily 25 km EASE-Grid 2.0, Version 1 [Data Set]. Boulder, Colorado USA. NASA National Snow and Ice Data Center Distributed Active Archive Center. https://doi.org/10.5067/MEASURES/CRYOSPHERE/nsidc-0530.001. Accessed 13 Sept 2023
Rosane O (2019, May 15) German far right party attacks 16-year-old greta thunberg ahead of EU Elections, EcoWatch. https://www.ecowatch.com/german-far-right-party-greta-thunberg-2637190985.html
Santer BD et al (1996) A search for human influences on the thermal structure of the atmosphere. Nature 382:39–46
Santer BD et al (2003) Contributions of anthropogenic and natural forcing to recent tropopause height changes. Science 301:479–483
Scheidel A et al (2020) Environmental conflicts and defenders: a global overview. Glob Environ Chang 63:102–104
Scruton R (2012) How to think seriously about the planet: the case for environmental conservatism. Oxford University Press
Serhan Y (2021) When the far right picks fights with a teen, Atlantic, August 14. https://www.theatlantic.com/international/archive/2021/08/greta-thunberg-far-right-climate/619748/

Sherman G (2014) The loudest voice in the room: how the brilliant, bombastic Roger Ailes built Fox news-and divided a country. Random House

Snyder CR, Train T, Schroeder LL, Pulvers KM, Adams V, Lamb L (2000, Summer) Teaching the hope recipe: setting goals, finding pathways to those goals, and getting motivated. Reaching Today's Youth 4(4):46–50

Telesca JE (2020) Red gold: the managed extinction of the Giant Bluefin Tuna. University of Minnesota Press, Minneapolis

The Washington Post (2019) No bold plans come from United Nations climate summit. The Washington Post, Dec 16. https://www.washingtonpost.com/lifestyle/kidspost/no-bold-plans-come-from-united-nations-climate-summit/2019/12/16/7b27b33c-1b70-11ea-87f7-f2e91143c60d_story.html

United Nations (2023) Climate change. Sustainable Development Goals. UN. https://www.un.org/sustainabledevelopment/climate-change/

University of East Anglia (2023) (n.d.) CRU temperature datasets. University of East Anglia https://crudata.uea.ac.uk/cru/data/temperature/

Varshalomidze T, Siddiqui U (2021) Leaders issue doomsday warning to tackle climate crisis. Aljazeera, Nov 1

Velicogna I et al (2020) Continuity of ice sheet mass loss in Greenland and Antarctica from the GRACE and GRACE follow-on missions. Geophys Res Lett 47(8):e2020GL087291

Von Schuckmann K, Cheng L et al (2020) Heat stored in the Earth system: where does the energy go? Earth System Science Data 12(3):2013–2041

Walker J, Heffner F (2010) Resilience as a critical factor in the workplace. The New Worker

Westerhold T et al (2020) An astronomically dated record of Earth's climate and its predictability over the last 66 million years. Science 369:1383–1387

Wolman BB (1989) Dictionary of behavioral science. Academic Press

Chapter 3
Managing Risks in the Agricultural Sector Facing Climate Change: Insights from Morocco

Fouad Elame, Youssef Chebli, Meriyem Koufan, Khalid Azim, Tarik Benabdelouahab, Ahmed Wifaya, Youssef Karra, Jamal Hallam, and Hayat Lionboui

Abstract Agriculture represents a central component of the economy of all African countries, as reported by many international organizations, representing an important source of income for a significant part of the world's population. In countries where rainfall is more uncertain due to climate change, risk management approaches would reduce agricultural risk and therefore the variability of farmers' incomes. This chapter is a review of literature that introduces the main concepts related to agricultural risks and highlights some important measures that have been adopted by the green generation plan in the Moroccan agricultural sector. To reduce disaster risk and build resilience in agriculture, a range of measures were taken by the Moroccan

F. Elame (✉) · M. Koufan · K. Azim · A. Wifaya · Y. Karra · J. Hallam
National Institute of Agronomic Research, Regional Center of Agadir, Agadir, Morocco
e-mail: fouad.elame@inra.ma

M. Koufan
e-mail: meriyem.koufan@inra.ma

K. Azim
e-mail: khalid.azim@inra.ma

A. Wifaya
e-mail: ahmed.wifaya@inra.ma

Y. Karra
e-mail: youssef.karra@inra.ma

J. Hallam
e-mail: jamal.hallam@inra.ma

Y. Chebli
National Institute of Agronomic Research, Regional Center of Tangier, Tangier, Morocco
e-mail: youssef.chebli@inra.ma

T. Benabdelouahab · H. Lionboui
National Institute of Agronomic Research, Regional Center of Rabat, Rabat, Morocco
e-mail: tarik.benabdelouahab@inra.ma

A. Sarkar et al. (eds.), *Risk, Uncertainty and Maladaptation to Climate Change*, Disaster Risk Reduction, https://doi.org/10.1007/978-981-99-9474-8_3

government. These measures focus on investing in technological innovation, climate-smart agriculture practices, and early warning systems. Improving access to resources such as land, water, inputs, and credit, and enhancing knowledge and skills through training and capacity building can also help to reduce vulnerabilities and increase resilience. Strengthening governance and policy frameworks, building partnerships and collaboration between the private and the public sectors, and promoting social protection mechanisms were also implemented by the government to reduce risk and building resilience in agriculture. Most of these actions will reduce the effect of certain risks and help promote the agricultural sector in the years to come.

Keywords Agriculture · Risk · Climate change · Resilience · Technological innovation

3.1 Introduction

The agricultural sector bears the impacts of climate change related disasters. Extreme weather events regularly affect agriculture, food security, water resources and health. This sector suffers 82% of the negative effects of the drought, which is a gradual consequence of climate change, compared to 18% for all the other sectors (FAO 2021).

Indeed, Agriculture is a sector that is highly vulnerable to a range of risks, including natural disasters, pests and diseases, market fluctuations, and socio-economic factors (Komarek et al. 2020). These risks can have a significant impact on farmers' livelihoods and food security, particularly in developing countries, where agriculture is a key driver of economic growth and poverty reduction. Policy and planning for risk reduction and management in agriculture is therefore crucial for ensuring sustainability of agricultural production and the well-being of farming communities (Kanchanaroek and Aslam 2018).

Optimizing policy and strategic planning for the mitigation and control of agricultural risks necessitates the active involvement of a diverse array of stakeholders. This collaborative effort encompasses governmental institutions, farmers, agricultural associations, and other pertinent entities. It requires a comprehensive understanding of the risks facing agriculture, as well as the socio-economic and environmental factors that influence these risks. Policy and planning for risk reduction and management in agriculture also requires a multi-disciplinary approach that draws on expertise from a range of fields, including agriculture, economics, environmental science, and social sciences (Meinke et al. 2009).

Critical elements within the framework of policy and planning for the reduction and management of risks in agriculture encompass the following: conducting risk assessments and mapping, formulating strategies for risk reduction and management, and executing measures for risk mitigation (Conant and Brewer 2022). Effective policy and planning also involves the improvement of governance structures that support risk reduction and management in agriculture, including the establishment of early warning systems, the provision of financial and technical support to farmers, and the promotion of knowledge-sharing and capacity-building.

Overall, policy and planning for risk reduction and management in agriculture is essential for building resilient and sustainable agricultural systems that can withstand the challenges and uncertainties of the future. Policymakers and stakeholders can play a pivotal role in safeguarding the sustainability of agriculture and the well-being of farming communities by adopting a proactive and interdisciplinary approach to risk management.

In this review, we will list the most important concepts related to the risk in agriculture. We will also introduce development strategies regarding risk reduction in the agricultural sector, by taking Morocco as a case study. We will also highlight the main strategies implemented by the Moroccan government in the African continent, as a case study in developing and implementing reforms and policies to mitigate climate change and minimize agricultural risks.

3.2 Disaster and Vulnerabilities in Agriculture

Disasters and vulnerabilities in agriculture can have significant consequences, both for food security and for the overall economy of a country (Parven et al. 2022). Agriculture stands as a pivotal sector, supporting the livelihoods of countless individuals across the globe. Disruptions triggered by disasters can exert profound consequences on both the accessibility and cost of food, in addition to affecting the earnings of farmers and rural communities in significant ways (Rahaman et al. 2021).

Disasters such as droughts, floods, storms, and wildfires can damage crops, contaminate water sources, and destroy infrastructure such as irrigation systems and storage facilities (Shreve and Kelman 2014). These events can lead to crop failure, reduced yields, and loss of livestock, resulting in food shortages, price hikes, and decreased income for farmers. Moreover, inherent vulnerabilities in agriculture systems can exacerbate the impacts of disasters (Singhal and Jha 2021). These vulnerabilities may include limited access to resources such as land, water, and inputs, insufficient knowledge and skills among farmers, inadequate infrastructure, and weak governance and policy frameworks. These factors can also diminish the resilience of agricultural systems in the face of shocks, thereby constraining the capacity of farmers and rural communities to recover and embark on a path of sustainable reconstruction.

3.2.1 Vulnerability

Vulnerability is a complex and highly interconnected concept with several other concepts like risk, adaptation, resilience, hazards, etc. Despite this conceptual complexity, vulnerability assessment is very useful in guiding the choices of policy makers.

Many similar definitions of vulnerability exist in the literature. Vulnerability means "the degree to which a unit at risk is likely to suffer from exposure to a disturbance or stress, and the ability (or inability) of the unit at risk to cope, get out of it or adapt in a fundamental way (by becoming a new system or by disappearing)" (Kasperson 2000 cited by Ogouwalé 2013). As per the Intergovernmental Panel on Climate Change (IPCC) report from 2001, vulnerability can be defined as "the extent to which a system is susceptible to deterioration or at risk of experiencing significant damage as a result of fluctuations and unpredictability in rainfall patterns; this is the case of food products which can deteriorate depending on excess or insufficient rainfall" (GIEC 2001).

According to Hewitt (1983), vulnerability originates from a lack of access to resources linked to poverty and the marginalization of individuals and populations. This vulnerability is manifested through behaviors and coping mechanisms to stress of the affected population. For Downing (1990), vulnerability constitutes a comparative assessment applied to either a population or a geographic area. It pertains to the fundamental factors that shape exposure to famine and the degree of susceptibility to the repercussions of famine. Brooks (2003) provides a definition of biophysical vulnerability, describing it as the extent of harm inflicted upon a system by a specific event or hazard, for example yield, loss or death toll (Tzilivakis et al. 2015). Based on the capability approach, Dubois and Rousseau (2008) in the field of social sciences consider that improving people's capabilities makes it possible to reduce vulnerability to the various risks encountered and they define vulnerability by the relationship between capability and risk (*Vulnérabilité* = *Capacité/Risque*).

The vulnerability assessment model is founded upon the IPCC definition of vulnerability, as outlined in the 2001 report, which comprises three primary elements: exposure, sensitivity, and adaptive capacity. Vulnerability, linked to climate, is the component of vulnerability that has been studied the most (Fig. 3.1). Vulnerability to climate change is frequently characterized as a function of the system's exposure to climate change (including its nature, magnitude, and rate of change), its sensitivity (entailing potential consequences), and its adaptive capacity. These components collectively encompass distinct aspects of vulnerability (GIEC 2001, 2007; Polsky et al. 2007; Tzilivakis et al. 2015; Fatemi et al. 2017).

For a comprehensive assessment of vulnerability to climate change and the identification of critical adaptation measures, it is essential to deepen our understanding of the biophysical and socio-economic factors that mitigate sensitivity to bolster the adaptive capacity of the systems in question. Several studies provide interesting classifications of vulnerability factors and indicators, with categories often linked to the dimensions of sustainable development, namely physical, economic, social,

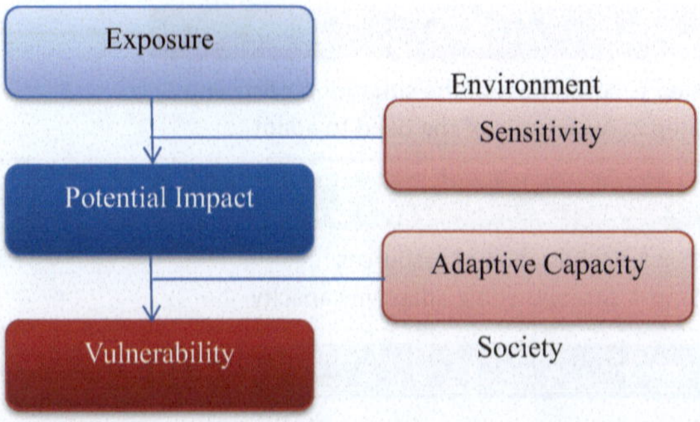

Fig. 3.1 Understanding vulnerability in the fourth ipcc report: conceptual framework

environmental, and institutional aspects of governance. For example, we can cite the United Nations Classification (United Nations 2004) which distinguishes groups of vulnerability factors:

i. Environmental and physical factors that describe the condition and status of the environment and the exposure of vulnerable elements within a particular geographic area.
ii. Economic factors, encompassing the financial resources of individuals, population segments, and communities.
iii. Social factors, encompassing non-economic aspects influencing the well-being of individuals, population groups, and communities. These include education levels, security, access to basic human rights, and the quality of governance.

Vulnerability is, therefore, an emerging interdisciplinary concept based on research on risks, hazards, and disasters. This concept incorporates qualitative and quantitative approaches that must be analyzed as a whole in order to implement sustainable adaptation approaches.

3.2.2 Adaptation

The future changes in the climate will profoundly affect the way our societies function. The problems linked to the means of adaptation are now at the heart of the major challenges of the years to come. Adaptation is a process by which communities and ecosystems adjust to changes and associated effects, in order to limit negative consequences and take advantage of potential benefits (Ouranos 2010).

The IPCC defines several different types of adaptation (IPCC 2007):

i. Anticipatory adaptation, which involves acting before the effects of anticipated changes are felt.

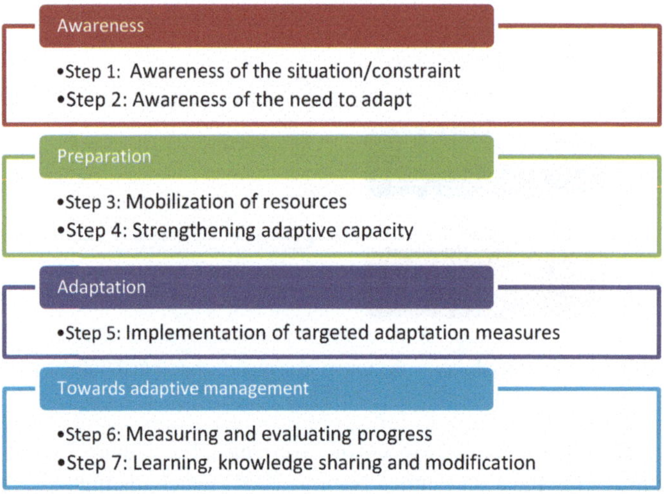

Fig. 3.2 The main steps of the adaptation process. *Source* Adapted by authors according to Eyzaguirre et Warren 2014

Fig. 3.3 Map of the study area

ii. Autonomous adaptation which consists of a spontaneous but unplanned response to climate impacts, which can arise from "ecological changes in natural systems or from an evolution of market conditions or the state of well-being in human systems."
iii. Planned adaptation, which is based on the definition of a strategy aimed at implementing measures to respond to observed and experienced climate variations.

Adaptation is a dynamic process involving the modification of both natural and human systems in response to observed or anticipated climatic stimuli, along with their corresponding effects and impacts. This entails a shift in methodologies, practices, and infrastructures, with the goal of mitigating potential harm or capitalizing on the opportunities presented by climate change. It induces adjustments to reduce the vulnerability of certain communities, regions, or activities/sectors for many periods, ranging from a few years to several decades.

According to Eyzaguirre and Warren (2014), the adaptation process encompasses 4 phases for a total of seven stages as follows:

i. *Awareness of climate change*: The adaptation process begins once the person or organization considers climate change to be a threat or an opportunity.
ii. *Awareness of the need to adapt*: Recognizing the magnitude of the problem makes it possible to consider the adoption of adaptation measures as a solution.
iii. *Mobilization of resources*: Awareness can lead people and organizations to devote human or financial resources to the problem, in order to help clarify the nature of the threats or opportunities.
iv. *Strengthening adaptive capacity*: The application of scientific data, financial resources, and skills to targeted activities such as the examination of issues, risk assessment and in-depth process analysis, makes it possible to acquire knowledge to support appropriate decisions.
v. *Implementation of targeted adaptation measures*: Concrete measures are implemented to reduce vulnerability (risk or exposure) to or to take advantage of opportunities that arise.
vi. *Measuring and evaluating progress*: Measuring and evaluating the effectiveness of adaptation measures, as well as associated assumptions and uncertainties, provides the insights needed to put in place improved management practices.
vii. *Learning, knowledge sharing and modification*: The last stage concerns the improvements made to the measures implemented and the transfer of lessons to future adaptation initiatives.

Based on the concepts introduced in this part of the chapter, we were able to show that improving knowledge, access to information and building skills through training and education is a key solution to reduce vulnerabilities and increasing resilience in agriculture (Figs. 3.2 and 3.3).

3.3 Risk Identification and Assessment in Agriculture

A risk is defined as any uncertain event occurring with a certain intensity and likely to reduce the company's performance (Declerck 2010). Risk is the likelihood that adverse consequences, damages, will actually materialize. According to Harwood et al. (1999), risk is an uncertainty to which the economic agent attaches importance and has an impact on his individual well-being, whereas uncertainty is a situation where we do not know what will happen. So, risk is necessarily uncertainty, but uncertainty is not necessarily risk.

For Moschini and Hennessy (2001), uncertainty represents several income possibilities associated with a chosen action, while risk is the result of a decision made under uncertainty. According to Winsen et al. (2011), risk is the result of a lack of information; it is objective and calculable. It can be negative or positive. By grouping the ideas of these authors, it is possible to conclude that uncertainty is linked to a lack of information in relation to the consequences of an event, while risk is the result of an uncertainty that can be measured.

The consequence of the risk can be positive, because for people in finance, the risk is remunerated, but it can also be negative, representing a loss. Some authors even orient their definition of risk towards adverse consequences, and thus, define it as income variability with an emphasis on loss. By analyzing risk in finance and insurance, Cordier (2008) manages to define risk as the harmful consequence of a random event. Risk is, therefore, inseparable from the harmful events it can cause. For Miller et al. (2004:1), risk can be a potential loss, but it can be a source of potential profit.

Risk identification in agriculture involves the process of identifying potential risks that may affect the agricultural production, profitability, and sustainability of the farm. This process involves evaluating the likelihood and impact of various risks, such as weather events, market fluctuations, and pest outbreaks, and developing strategies to mitigate or manage these risks. To identify risks in agriculture, farmers need to consider a wide range of factors, including weather patterns, soil conditions, crop selection, pests and diseases, market prices, and regulatory requirements.

In agricultural activity, by nature dependent on climatic and sanitary conditions, the risk is particularly present. This notion can ultimately be summed up as the harmful consequence of an unwanted event (Cordier et al. 2008). These are risks related to the possibility that results will differ from reasonable expectations. These differences can be attributed to climatic conditions, parasites, pests, technology, or management. Agricultural risk means any risk that may affect agricultural production or the performance of fieldwork.

Some authors classify risks into two categories (Antón et al. 2011; Bauer and Bushe 1994; Miller et al. 2004), distinguishing them by production risks and market risks, business and financial risks, or risks related to the farm and those related to the world outside it. Others go as far as to classify them into seven categories (Holzmann and Jorgensen 2001). Nevertheless, much like the definition of agricultural risk, several commonalities emerge in how to classify agricultural risk as follows:

i. Production risk is a risk related to yields, weather conditions, pests, and diseases. Some authors include other notions within production risk, including Moschini et al. (2001) and Cordier et al. (2008) with the notion of product quality. For Cordier, the production risk is called the climatic and health risk, while for other authors, these are the so-called natural risks.
ii. Market risk is the risk associated with fluctuations in the prices of outputs as well as agricultural inputs (Cordier et al. 2008:38; Hardaker et al. 2004:6; OCDE 2010:23–24). This type of risk is called by some authors marketing risk according to Baquet et al. (cited by OCDE 2010), or price uncertainty by Moschini et al. (2001).
iii. Financial risk is a risk linked to the borrowing capacity of operators and their ability to honor interest. An increase in interest rates or an increase in the debt ratio therefore increases the financial risk of the company. For Hardaker et al. (2004) and Cordier (2008), the demand for unforeseen credit repayment, the variation of the exchange rate and the possibility of borrowing also affect the financial risk.
iv. Institutional and legal risk is the risk associated with public authorities. Authors such as Hardaker et al. (2004) and Moschini et al. (2001) set out the notions of state payments, subsidies, and taxation, while authors such as Musser and Patrick as well as Baquet et al. (OCDE 2010), rather speak of legal and environmental risks linked to legal proceedings. Hardaker et al. (2004) think that institutional risk overlaps with several other types of risk, including so-called "sovereign" political risk. This is the risk that some countries will not honor their trade agreements. There is also the risk of a business relationship, in connection with a business partner or an organization. Cordier et al. (2008) insist instead on the risk associated with changes in agricultural policies or regulations.
v. Human risk is linked to everything that directly or indirectly affects individuals and their social life. It is only mentioned by a few authors. Hardaker et al. (2004) emphasize harmful events such as death, divorce, and disease. Musser and Patrick as well as Baquet et al. (cited by OCDE 2010) rather refer to the risk of human resources, which are events related to family or external labor preventing them from working. Cordier et al. (2008) broaden the definition of this risk by including adverse human events such as breakage, theft, and destruction of equipment.

It is important to know the real and potential consequences of these impacts and risks on ecosystems. Indeed, knowledge of the current state of the environment and the related problems is essential and should constitute a prerequisite for decision-making processes and operational mechanisms in terms of preservation and sustainable management of the environment.

3.4 Development of Risk Reduction and Management Strategies

Risk management is the systemic and practical managerial approach to limiting potential damage and loss. Risk management includes the assessment of risks and their analysis, as well as the implementation of specific strategies and actions to control, reduce and transfer them (Hardaker et al. 2004; Corider et al. 2008; UNISDR 2009; Theuvsen 2013). Risk management aids in reducing the probability of occurrence of the event considered, in setting up systems aimed at reducing the harmful consequences. It is necessary to intervene and make producers understand the risks in order to facilitate sustainable management of these phenomena. In agriculture, risk-related losses can be very significant as it makes risk management in the conduct of an agricultural business crucial.

Managing risk process, as outlined by Theuvsen (2013), encompasses four main steps. Firstly, in the risk identification phase, the goal is to categorize the most relevant types of risk for a specific farm. It is important to note that risks faced by farmers may vary significantly from those encountered by breeders due to differences in the types of threats. For instance, plant-related risks are distinct from animal-related risks due to variations in the parasites that affect these different systems. Moving on to the risk assessment stage, the key question here revolves around determining the level of attention a particular risk demands. To assess this, two criteria come into play: (i) the objective measurement of the risk's probability of occurrence, which can be based on empirical data (e.g., rainfall) or subjectively evaluated, and (ii) the assessment of potential losses associated with the occurrence of the risk. Subsequently, the risk management step involves considering various strategies to address identified risks. These strategies include avoidance, which involves shifting activities away from those most exposed to a specific risk. Risk transfer through mechanisms like insurance, risk mitigation through diversification of farm activities, and risk acceptance, is suitable when the risk's probability and associated losses are low. Lastly, the risk control phase aims to reduce the occurrence of the risk to a threshold deemed acceptable, enhancing the farm's resilience and stability.

Although there are several risk management models, a comprehensive approach to risk management seems more suitable for many cases, knowing that risk does not have linear interactions, but rather multidimensional ones that must be analyzed as a whole and not individually. The global approach proposed by the OCDE (2010) as given in Table 3.1 classifies risk management strategies according to their scope of action in columns (household, market, and public authorities), and in rows according to the three types of strategies (reduction, mitigation, and adjustment). According to this approach, risk reduction reduces the likelihood of an adverse event, while risk mitigation minimizes the potential impact of an adverse event.

This classification, developed initially by Hardaker et al., 2004, encompasses several levels of actors, including farm or household, community, market, and government. In what follows, we will focus mainly on the strategies maintained at the agricultural level.

Table 3.1 Risk management strategies according to the types of strategies and actions

Risk management	Farm/household/community	Market	Public authorities
Reduction of risks	Technological choices	Training in risk management	• Macroeconomic policies • Disaster prevention • Prevention from animal diseases
Mitigation of risks	Diversifying production and sharecropping	• Contracts and Options • Assurance • Vertical integration • Production and • Marketing contracts • Sales spread • Diversification of financial investments • Non-agricultural activity	• Income-smoothing tax regime • Countercyclical programs • Border and other measures in case of epizootics
Adjustment to risk	Loans from neighbors and family members mutual aid	• Sale of financial assets • Savings and borrowing from banks • Non-farm income	• Disaster payments • Social Help • Agricultural support programs

Source Based on OCDE (2010)

Technological choices can influence yield, income, and their variability, such as the choice of a more intensive mode of livestock farming, which means that you are no longer exposed to climate risk, unlike grazing (Hardaker et al. 2004). Other technologies make it possible to reduce the effect of certain risks, such as irrigation to compensate for the lack of precipitation.

Diversification is a risk management strategy expressed through a variety of production, location of production and sources of income. This method targets production, market, and financial risks. The diversification of production makes it possible to limit the consequences of a harmful event, such as diseases, pests, etc. (Hardaker et al. 2004:273). In addition to this advantage, there is the improvement of yields, a more sustainable crop rotation and a spreading of demand for labor, machinery, and remuneration over time.

Sales contracts are a category of risk management strategy rather than a single strategy. The contracts include the sale by a cooperative, the contracting of inputs or outputs, hedging contracts on the futures and options for markets. This set of risk management strategies targets market risk. The first sales contract-type risk management strategy is called price pooling (Hardaker et al. 2004). The principle consists in organizing the producers under a single entity to obtain an average price of sales, in order to fight against short-term price fluctuations. The second sales contract-type risk management strategy is called "contracting" consisting of entering into a contract between a seller and a buyer in order to determine the price in advance. The third contract-type risk management strategy involves hedging contracts in futures markets. This tool consists of selling and buying delivery contracts, so as not to be a prisoner of the price obtained during the harvest. The fourth contract-type risk management strategy in this category is the option which is very similar to "hedging" except that the producer is not obliged to sell or buy. The farmer pays a premium to have the right to sell or buy at a certain time, without being forced to do so, all within a time defined when buying the option. This allows the farmer to take advantage of a market opportunity while securing the price of his production.

Various types of insurance are offered to farmers, both to cover their production against certain weather vagaries and to cover risks of human nature, such as fires, injuries, and death. The concept of insurance is that the insured concedes the risk to a third party, such as an insurance company, in exchange for a risk premium. The insurance company uses the principle of pooling. The farmer's choice for insurance is a function of the insurance premium paid in relation to the possible damage caused and his aversion to risk.

Financial leverage consists of changing the financial structure of the company in order to increase the return on stockholders' equity. In this case, the optimal choice of the combination between stockholders' equity and debt depends on the risk aversion of the farmer.

Flexibility is a risk management strategy to adapt to events by maintaining or increasing the farmer's options to react to bad events or to benefit from opportunities. In the agricultural sector, this tool applies to different aspects of the business: assets, products, markets, costs and working time.

In addition to these different strategies, there are also the various agricultural aid programs and the intervention of the public authorities, whether in terms of prevention against natural disasters (fires, floods, diseases, etc.) or social aids, compensation premiums and tools for intervening in border prices, taxes, and fiscal charges.

The agricultural policies undertaken in Morocco since the end of the 1980s have continued to make profound modifications to the instruments of previous periods both in terms of orientation and measures with a view to resolving different types of agricultural issues. Many strategies facing climate change and risks in agriculture were applied at the farm and at the market level. However, some constraints remain and are related, in particular, to a governance deficit (public institutions weakness at the territorial and regional level, weaknesses in terms of public–private partnership and professional organization, increased centralization and up-to-down public action (DEPF 2019). The following section will be dedicated to discussing Moroccan agricultural strategies regarding agricultural risk.

3.5 Planning for Resilience in the Moroccan Agriculture

The impact of climate change is expected to be particularly significant in the agricultural sector in Morocco because of the position of the agricultural sector for the economy due to the predominance of the areas dependent on rainfall and the low adaptive capacities. Indeed, the sensitivity to climate change, which represents the degree to which the agricultural sector will be affected, depends on several elements. The importance of rainfed agriculture (95% of the total agricultural area) and water withdrawals for agricultural activities (87% of total amount) are decisive for the evolution of productivity. Sensitivity also takes into account the economic importance of the agricultural sector, which contributes on average 17% of GDP, considered very high in Morocco compared to other Maghreb countries, and nearly 45% of employment (MAPMDREF 2021).

Climatic changes and projections for Morocco highlight a rise in temperatures and a drop in precipitation which result in a northward progression of arid and semi-arid climates (Woillez 2019). Precipitation is experiencing a marked decline, particularly in the north of the country where it reaches between -10 and -25/mm per decade for the period 1951–2010. However, the main characteristic remains the spatial and temporal variability. The decrease in average annual precipitation is about 20% between 1960 and 2005 but it is not significant in all the locations studied. Similarly, the trends relating to the intensity of precipitation and the duration of wet periods are heterogeneous, with a predominance of the frequency of intense rains.

Regional climate model projections indicate a widespread temperature increase, especially in the eastern region of the country, along with an overall decline in precipitation of 15% by the year 2050. Nevertheless, there will be notable year-to-year variations in precipitation. This rise in temperatures, coupled with increased evapotranspiration due to reduced precipitation, is anticipated to exert greater pressure on water resources. Projections from hydrological models identify Morocco as a significant focal point for water stress, as noted in studies by Wolliez (2019) and Filahi et al. (2017). Consequently, changes in diurnal thermal amplitudes combined with irregular interannual rainfall trends transform the seasonality and quality of agropastoral areas and influence agricultural calendars and technical itineraries, which has immediate repercussions on yields and labour costs and may involve decreases in productivity with a potential duration of longer occupation of the areas by the production systems.

According to Elame et al. (2020), climate projections scenarios for the next 15 years show that irrigated crops, less directly impacted by climate change, could see their yields drop or even witness a conversion of irrigated lands into rainfed lands in certain agricultural areas. The adaptive capacities of irrigated agriculture, in particular the improvement of the water use efficiency and the support of farmers to adopt production technologies that save water therefore appear crucial.

3.6 Moroccan Strategies Regarding Agricultural Risks

It is well known that the most vulnerable populations are mostly poor and live in rural areas. These populations survive through natural resources usage. Any changes to the environment due to climate variability affect their livelihoods and increase their vulnerability. One of the consequences of these changes is the rural exodus of young people increasing in rural urban migration, overcrowding the cities. Thus, it becomes important to increase the adaptation capacities of rural communities depending on agriculture by building capacities, raising awareness of climate variability, improving their knowledge of resources management, and creating alternative income-generating activities.

The agricultural sector in Morocco employs a range of adaptation strategies, including the cultivation of drought-tolerant crop varieties, the reduction of vegetative cycle duration, the adoption of early-maturing crop varieties, and the implementation of flood-recession cultivation in watersheds. Additionally, strategies encompass the promotion of irrigated agriculture with controlled water usage, the adoption of greenhouse cultivation practices, and the pursuit of sustainable agricultural intensification. These measures are particularly pertinent in response to the observed trend of a shortened rainy season. With regard to agro-sylvo-pastoral systems, management practices are modified according to climatic realities. The participatory management of natural rangelands and the association between agriculture and livestock enhance the resilience capacity of populations in an unfavorable climatic context. Livestock is generally chosen according to their resistance capacities. For example, camels and goats are preferred in arid areas because of their lower need for water and fodder.

Likewise, due to the reduction in rainfall, the drop in groundwater levels and the modification of surface water flows, in recent years, new local practices can be adopted such as control and storage of rainwater in storage basins, retaining dikes, surface runoff slowdown thresholds, traditional wells, etc.

Another common strategy to mitigate climatic hazards involves diversifying economic sectors beyond agricultural production, thus reducing dependency on the unpredictability of harvests. In addition to the risks related to climate change, other risks seem important for farmers such as animal and plant health risks, market risks and human risks. Health risks, such as illness and mortality, result in lost yield and increased costs. Market risk includes the fluctuation of the prices of inputs and outputs having effects such as a decrease in income and a slowdown in investments. Indeed, with the outbreak of the Covid-19, most agriculture input prices have increased considerably. In addition, with the war in Ukraine, the world economy felt the effects of increasing inflation and growth decreasing. This resulted in a rise in the prices of raw materials, such as food and energy, which further increased inflation and had the direct effect of reducing the value of income.

The main adaptation strategies are both technical (cultivation practices, varieties, etc.) and economic. Better valuation of agricultural production by improvement of marketing by integrating producers into short circuits or better information on prices could improve adaptation capacities. Other strategies to manage these risks are futures contracts, learning about market prices, diversification, and grain storage. Government intervention through production subsidies is also necessary in this kind of situation.

In this context, the Moroccan government has launched many important strategies. From structural adjustment policies in the eighties, through the Green Morocco 2008–2020 plan, to the Green Generation 2020–2030 plan, several strategies have marked Moroccan agriculture and allowed larger opening to the rest of the world through trade liberalization. Among the risk-related measures, financial and banking measures are worth mentioning. These include facilities granted for investments like the "Livestock Protection" Fund, the price subsidy for agricultural products and livestock feed, Fund to Combat the Effects of Natural Disasters (FLCN) in 2009. All these investments are aimed at reducing the risks that could weigh on agriculture as well as other related sectors.

Farm machinery is also an important element in field crop production and some risk management strategies target this element. For farmers, the technological choice applies to two subjects: selected seeds to manage production risks, and machinery to simplify the farmer's work, minimize labor, and input losses in the field. Investing in early warning systems is also one of the measures to be advocated in order to help decision-makers and farmers to take the necessary measures to reduce agricultural risks. Indeed, in 2013, Morocco established a national system known as "CGMS-MOROCCO" (Crop Growth Monitoring System-Morocco) to monitor the agricultural campaign and provide agro-meteorological predictions for cereal harvests. Information is a crucial element in the field of risk. For some authors, information distinguishes risk from uncertainty. It is therefore quite appropriate to find it among the elements influencing the perception of risk.

It is thus evident that with the availability of more information to the stakeholders there is less level of uncertainty. With more certainty, it is easier to foresee the different eventualities and therefore have control of the situation. The "Green generation" plan 2020–2030 advocates information and data based on technological innovation. One of the central objectives of the Green Generation Plan is to facilitate the emergence of a new generation within the middle-class population that depend on agriculture for their livelihood. This goal is to be achieved through the habilitation of one million hectares of collective land, the creation of 350,000 employment opportunities, and the integration of at least two million farmers into digital service platforms, as outlined in the Agricultural Development Agency Report of 2020 (ADA 2020). The development of innovation platforms is one of the main aspects that have been promoted by this strategy in order to connect farmers to the market chain supply.

The strengthening of adaptive capacities also appears to be decisive, particularly with regard to rainfed crops, to deal with the reduction in water resources and changes in areas favorable to cereal cultivation. Soil conservation and fertility preservation techniques (sowing under plant cover for example, agroforestry, agroecology) could see their increased interest in this context. Indeed, the Moroccan green generation plan has set a target of one million hectares in direct seeding. Sowing experiments under plant cover in certain areas in Morocco have shown that yields were improved in dry years. The rise of cultivation areas favorable to cereal crops towards mountain areas shows the interest of agroforestry in reducing soil erosion (ACCAGRIMAG 2018). The diversification of crops, the selection of adapted varieties and the development of agro-ecological practices could also reduce the risks of the negative impacts of climate change.

Complementary strategies for public and private partnership and social protection therefore seem crucial. In fact, Social protection aims to reduce the vulnerability of individuals to these hazards. This component has been intensively promoted by the Ministry of agriculture and the "*Crédit Agricole*" Bank. It can take the form of a cash or in-kind benefit and generally comprises three main components, namely, social assistance, social insurance, and labor market programs. Accelerating the implementation of the insurance coverage mechanism in the agricultural sector was an important element of the "Morocco green plan" strategy.

Indeed, a national agricultural risk management framework has been developed by the Ministry of Agriculture and Maritime Fisheries, to move from ex-post crisis management to risk mitigation, ex-ante investments and new insurance products. This strategy aims to reduce the vulnerability of small farmers to agricultural risks, promote and secure agricultural investment and provide direct public support for insurance products for better management of agricultural risks (MAPM 2010).

Government-subsidized agricultural insurance programs to cover perishable crops like olives, fruits, and vegetables. Multi-risk harvest insurance covers four types of cereals and five types of legumes against six different types of risks: drought, flooding, hail, frost, strong winds and sandstorms. Multi-risk climatic guarantee program for fruit trees includes Rosaceae, Citrus and Olive trees against the same types of risks mentioned above. The main products offered by private insurance covers agricultural machinery, farm liability, agricultural fires, and livestock mortality. Also, investments aiming at increasing yield or increasing production were considered in order to develop farms in the medium term. In addition, an innovative insurance scheme to cover the consequences of catastrophic events was adopted in 2018. The law introduced a private insurance scheme covering nearly 9 million people and created a Solidarity Fund against Catastrophic Events (FSEC), distinct from the FLCN.

From a global perspective, Morocco is relatively efficient in terms of technology adoption, particularly in large, irrigated areas. On the other hand, it appears to have weak capacities for institutional adaptation. The institutional environment does not make it possible to contribute significantly to the sustainable management of water resources or to determine priorities in adaptation measures (OCDE 2016). Therefore, the country must focus on inter-institutional collaboration and adopt a systemic multi-risk and multi-sectoral approach to risk management for better integrated responses to reduce climate change impacts in general and agricultural risks in particular.

3.7 Conclusion

Agricultural activity is conducted in a context where uncertainty and insecurity still exist, although to varying degrees. The causes are varied and related to the multiplicity of factors and conditions that interfere in the process of agricultural production. According to several international reports, climatic hazards added to the crisis of war conflicts do not only result in loss of human life, but they also cause the disappearance of agricultural means of subsistence and cause severe economic damage to households, countries, and regions, which may perish in the long term.

This chapter has two important contributions. First, it adds knowledge to the conceptual understanding of risk analysis in agriculture and methodological elements to apprehend decision-making strategies. Second, it describes the vulnerable state of the Moroccan agriculture and discusses the main management strategies and government interventions to reduce risk and increase the resilience of the sector. Among the key strategies adopted by the government in this direction is the "green generation plan" by investing in new technologies, in particular. Integrated responses and cross-sectoral collaboration are essential in disaster response. The country must capitalize on past experiences and adopt a systemic multi-risk and multi-sectoral approach to risk management, in order to anticipate the risks of disasters in agriculture and reduce their possible impacts.

References

ACCAGRIMAG (2018) Projet d'adaptation au changement climatique de l'agriculture au Maghreb (ACCAGRIMAG). Composante 2 : accroissement des capacités de résilience des systèmes de cultures pluviales par l'expérimentation et la diffusion de technologies innovantes d'adaptation au changement climatique

ADA (2020) Rapport de la Réunion régionale sur la nouvelle stratégie "Génération Green 2020–2030". Agence de Développement Agricole

Antón J, Kimura S, Martini R (2011) La gestion des risques agricoles au Canada. Éditions OCDE, 93 p

Baquet A, Hambleton R, Jose D, Eliason J (1997) Introduction to Risk Management

Bauer L, Bushe D (1994) Designing Risk Management Strategies: Faculty of Extension, University of Alberta

Brooks N (2003) Vulnerability, risk and adaptation: a conceptual framework. Norwich (UK): Tyndall Centre for Climate Change Research. 16 p. (Working Paper, n. 38)

Conant A, Brewer G (2022) Principles and practice: Towards disaster risk reduction in New South Wales, Australia. Urban Governance 2:285–295

Cordier J, Erhel AA, Pindard A, Courleux F (2008). La gestion des risques en agriculture de la théorie à la mise en oeuvre : éléments de réflexion pour l'action publique. Notes et études économiques(30), pp 33–71

Declerck F (2010) Multiplicité et ampleur des risques en agriculture: essai de typologie. Article scientifique, Edition la France agricole, 2 p

DEPF (2019) Le secteur agricole marocain : Tendances structurelles, enjeux et perspectives de développement

Downing TE (1990) Assessing socioeconomic vulnerability to famine: frameworks, concepts, and applications. Washington (Etats-Unis): U.S. Agency for International Development. 196 p. (FEWS Working Paper, n. 2.1). http://pdf.usaid.gov/pdf_docs/PNABJ875.pdf

Dubois JL, Rousseau S (2008) Reinforcing households' capabilities as a way to reduce vulnerability and prevent poverty in equitable terms. In Comim F, Qizilbash M, Alkire S The capability approach: concepts, measures and applications. Cambridge (UK): Cambridge University Press. pp 421–436. https://doi.org/10.1017/cbo9780511492587.014

Elame F, Doukkali R, Lionboui H (2020) Dynamic Modeling of Climate Change Impact on Agricultural Lands and Water Resources. In: Leal Filho W., Luetz J., Ayal D. (eds) Handbook of Climate Change Management. Springer, Cham. https://doi.org/10.1007/978-3-030-22759-3_41-1

Eyzaguirre J, Warren F (2014) Adaptation : établir un lien entre la recherche et la pratique. In: Warren F.J., Lemmen D.S. Vivre avec les changements climatiques au Canada: perspectives des secteurs relatives aux impacts et à l'adaptation. Ottawa (Ontario): Gouvernement du Canada.

FAO (2021) The impact of disasters and crises on agriculture and food security: 2021. Rome. ISBN 978–92–5–134071–4. https://doi.org/10.4060/cb3673en

Fatemi F, Ardalan A, Aguirre B, Mansouri N, Mohammadfam I (2017) Social vulnerability indicators in disasters: findings from a systematic review. International Journal of Disaster Risk Reduction 22:219–227. https://doi.org/10.1016/j.ijdrr.2016.09.006

Filahi S, Tramblay Y, Mouhir L, Diaconescu EP (2017) Projected changes in temperature and precipitation indices in Morocco from high-resolution regional climate models. International Journal of Climatology, 37(14), 4846–4863. Cité par Wolliez (2019)

Füssel HM (2007) Vulnerability: a generally applicable conceptual framework for climate change research. Glob Environ Chang 17(2):155–167. https://doi.org/10.1016/j.gloenvcha.2006.05.002

GIEC (2001) Bilan 2001 des changements climatiques : conséquences, adaptation et vulnérabilité. Genève (Suisse) : GIEC. 101 p. http://www.observatoireclimat-hautsdefrance.org/Les-ressources/Ressources-documentaires/Bilan-2001-des-changements-climatiques-Consequences-adaptation-et-vulnerabilite

GIEC (2007) Bilan 2007 des changements climatiques : rapport de synthèse Genève (Suisse) : GIEC. 103 p. https://www.ipcc.ch/site/assets/uploads/2018/02/ar4_syr_fr.pdf

Holzmann R, Jorgensen S (2001) Social Risk Management: A New Conceptual Framework for Social Protection, and Beyond. Int Tax Public Financ 8(4):529–556

Hardaker JB, Huirne RBM, Anderson JR, Lien G (2004) Coping with risk in agriculture

Harwood JL, Heifner R, Coble K, Perry J, Somwaru A (1999) Managing risk in farming: concepts, research, and analysis : US Department of Agriculture, Economic Research Service

Hewitt K (1983) Interpretations of calamity: from the viewpoint of human ecology. Boston (Etats-Unis): Allen and Unwin. 304 p. (The risks & hazards series, n. 1). ISSN 0261–0507

Kasperson JX, Kasperson RE (2001) A Workshop Summary – International Workshop on Vulnerability and Global Environmental Change. SEI Workshop, 39 17–19 May 2001, Stockholm Environment Institute, Stockholm, Sweden. http://www.sei.org

Komarek AM, De Pinto A, Smith VH (2020) A review of types of risks in agriculture: What we know and what we need to know. Agricultural Systems, No 178:1–10

Kanchanaroek Y, Aslam U (2018) Policy schemes for the transition to sustainable agriculture—Farmer preferences and spatial heterogeneity in northern Thailand. Land Use Policy 78:227–235

MAPMDREF (2021) Analyse de vulnérabilité aux changements climatiques. Ministère de l'Agriculture, de la Pêche maritime, du Développement rural et des Eaux et forêts www.agriculture.gov.ma

MAPM (2010) Ministère de l'agriculture et de la pêche maritime. Etude sur la gestion des risques et la mise en place d'un système d'assurance agricole au Maroc. Identification et hiérarchisation des risques agricoles

Meinke H, Howden SM, Struik PC, Nelson R, Rodriguez D, Chapman SC (2009) Adaptation science for agriculture and natural resource management—urgency and theoretical basis. Current Opinion in Environmental Sustainability 1:69–76

Miller A, Dobbins C, Pritchett J, Boehlje M, Ehmke C (2004) Risk management for farmers. Staff paper, pp 4–11

Moschin G, Hennessy DA (2001) Uncertainty, risk aversion, and risk management for agricultural producers. Handbook of Agricultural Economics, Volumes 1a et 1b, 18, pp 87–153

Moss RH, Brenkert AL, Malone EL (2001) Vulnerability to climate change: a quantitative approach. Richland (Etats-Unis) : Pacific Northwest National Laboratories. 70 p.

OCDE (2010) Gestion des risques dans l'agriculture : une approche holistique. Editions OCDE, Paris. 207 p. https://doi.org/10.1787/9789264075337-fr

OCDE (2016) Étude de l'OCDE sur la Gestion des Risques Maroc Principaux résultats, https://www.oecd.org/fr/gov/risques/gestion-des-risques-maroc-principauxresultats.pdf

Ogouwalé E (2013) Techniques d'évaluation de la vulnérabilité et de l'adaptation des systèmes humains et naturels aux risques climatiques, support du cours, 37p

Ouranos (2010) Élaborer un plan d'adaptation aux changements climatiques. Guide destiné au milieu municipal québécois. [Rapport], 48 p

Parven A, Pal I, Witayangkurn A, Pramanik M, Nagai M, Miyazaki H, Wuthisakkaroon C (2022) Impacts of disaster and land-use change on food security and adaptation: Evidence from the delta community in Bangladesh. International Journal of Disaster Risk Reduction 78:103119

Polsky C, Neff R, Yarnal B (2007) Building comparable global change vulnerability assessments: the vulnerability scoping diagram. Global Environmental Change, 2007/08/01, vol. 17, n. 3, pp. 472–485. https://doi.org/10.1016/j.gloenvcha.2007.01.005

Rahaman A, Kumari A, Zeng X-A, Khalifa I, Farooq MA, Singh N, Ali S, Alee M, Aadil RM (2021) The increasing hunger concern and current need in the development of sustainable food security in the developing countries. Trends Food Sci Technol 113:423–429

Shreve CM, Kelman I (2014) Does mitigation save? Reviewing cost-benefit analyses of disaster risk reduction. International Journal of Disaster Risk Reduction 10:213–235

Singhal A, Jha SK (2021) Can the approach of vulnerability assessment facilitate identification of suitable adaptation models for risk reduction? International Journal of Disaster Risk Reduction 63:102469

Theuvsen L (2013) Problems of World Ariculture. Volume 13 (XXVIII) Number 4. Scientific Journal, Warsaw University of Life Sciences press – SGGW.

Tzilivakis J, Warner DJ, Green A, Lewis KA (2015) Adapting to climate change: assessing the vulnerability of ecosystem services in Europe in the context of rural development. Mitig Adapt Strat Glob Change 20(4):547–572

United Nations. (2004) Living with risk: a global review of disaster reduction initiatives. Geneve (Suisse): United Nations. 429 p. https://www.undrr.org/publication/living-risk-global-review-disaster-reduction-initiatives

UNISDR (2009) The United Nations International Strategy for Disaster Reduction. UNISDR Terminology on Disaster Risk Reduction

Winsen FV, Wauters E, Lauwers L, Mey YD, Passel SV, Vaucauteren M (2011) Combining risk perception and risk attitude: A comprehensive individual risk behaviour model. Document présenté à EAAE 2011 Congress Change and Uncertainty, Zurich, Suisse

Woillez M-N (2019) Revue de littérature sur le changement climatique au Maroc. AFD, Research paper, no 2019–108

Chapter 4
Climate Change Adaptation, Risk Reduction and Indigenous Knowledge Based Resilience: A Case of Bonda Tribal Women in Odisha

Subrata S. Satapathy

Abstract Indigenous or tribal peoples' institutions, rights, and privileges often go unacknowledged. Additionally, Climate Change makes gender inequity, a major contributor to the suffering of women from indigenous population. Though the tribal women contribute significantly to both traditional and non-traditional forms of employment, with unpaid caregiving, and maintaining food security, they frequently experience prejudice both within and outside of their communities. The present chapter is a case study of the vulnerable tribal group in Odisha, the Bonda. The study aims to document the best practices carried out by the tribal women of this indigenous group. It focuses on the ways employed by them to preserve natural resources and the techniques they employ to mitigate the Climate Change challenges. The chapter is based on qualitative research. The findings are derived from primary survey conducted in Malkangiri district of Odisha. Data has been collected from interviews and focus group discussions with the Bonda tribal women. This study shows that these women have employed their indigenous knowledge in reverting the precarious consequences of climate change and steered the environment protection towards their favour. Their farming methods have evolved with the natural world.

Keywords Bonda women · Climate change · Indigenous · Tribes · Resilience

4.1 Introduction

Throughout history, indigenous communities around the world have developed unique and sustainable ways of interacting with their environments. These practices are deeply rooted in cultural traditions, wisdom, and hence, there exists an inherent understanding of the delicate balance between humans and nature. One significant aspect of indigenous environmental practices is the central role that women often

S. S. Satapathy (✉)
Christ Academy Institute of Law, Bengaluru, India
e-mail: subrata.s@calaw.in

A. Sarkar et al. (eds.), *Risk, Uncertainty and Maladaptation to Climate Change*,
Disaster Risk Reduction, https://doi.org/10.1007/978-981-99-9474-8_4

play. In many indigenous societies, women have traditionally been the primary caregivers for families and communities, and this responsibility extends to the care of the land and natural resources. Their connection to the land is not solely utilitarian but also spiritual, as it is intricately tied to their cultural identity and beliefs.

Indigenous women are often the keepers of traditional ecological knowledge, passed down through generations. They possess an intimate understanding of local ecosystems, medicinal plants, sustainable farming techniques, and natural resource management. This knowledge is crucial for maintaining biodiversity, ensuring food security, and mitigating the impacts of climate change. Furthermore, women in indigenous communities frequently engage in practices that promote ecological sustainability. These may include seed-saving, agroforestry, water conservation, and the preservation of traditional farming methods. Their respect for the land and its resources is often reflected in the emphasis on responsible resource extraction, rotational farming, and the minimization of waste. Traditional wisdom has a history of saving lives. The Moken, a nomadic tribal community that travels the seas of southern Thailand and Myanmar, noticed a dramatic retreat in the water on the coasts of Yan Chiak, Myanmar in 2004, just before the Indian Ocean tsunami struck. When the disaster struck, the entire hamlet relocated to a higher location, sparing many lives. The Moken survived while many others perished, according to the UNESCO Regional Advisor for Culture in Asia (UNESCO, Intangible Cultural Heritage 1992–2023), which highlights certain lessons to be gleaned from traditional, indigenous knowledge. In the meantime, it was discovered that conventional building materials, such as bamboo and thatch, utilised for dwelling construction near the sea based on conventional criteria, would not kill the people in the event of a collapse. As per a report published by United Nations International Strategy for Disaster Reduction, Simeulue, an Indonesian settlement of over 80,500 people, utilised their traditional knowledge to evacuate to adjacent hills during the tsunami, saving tens of thousands of lives (UNISDR 2006). Importantly, indigenous women's involvement in decision-making is essential for eradicating gender inequity and prejudice against them. Greater awareness must be given to the labour situation and economic contributions of indigenous women because they play a significant role in securing livelihoods and incomes. According to a research conducted by the ILO in Peru, indigenous women have particularly difficult access to financial and technical support, which hinders their efforts to establish businesses, connect to markets, and engage in activities like trading and producing handmade goods (ILO 2016).

However, the management of all agricultural output and natural resources is getting riskier by the day. Numerous women farmers in tribal communities rely on the same (IAASTD 2009). They are searching for methods to lessen the effects of global warming as well as successful adaption tactics for the anticipated repercussions of unpredictable rainfall, drought, and other problems. However, the success of the tribal people's sustainable practises is largely dependent on the goodwill among neighbours and the ability of the entire society to be resilient with outside pressures like privatisation and globalisation. Gender relationships are crucial to both. The main objective of this chapter is to comprehend and decode the gender-based effects of climate change on the Bonda tribal communities in Odisha. It explores the indigenous knowledge and practices employed by Bonda tribal women in mitigating climate change issues and to understand the resilience mechanisms towards the ever-enveloping climate change problem. The chapter also elaborates on the effects that are aggravated by structural changes in tribals' socio-economic systems, especially due to the recent impacts of privatisation, hurried and unsystematic development agenda of the government and gender-based- roles within tribal communities.

4.2 Research Methodology

The paper is based on qualitative research, supported by both fieldwork and secondary sources. The data has been collected through intensive field work in Malkangiri district of Odisha. Malkangiri's geographical isolation has contributed to the preservation of the tribe's cultural practices, though it has also presented them with challenges in terms of connectivity and access to modern amenities. The method of data collection was both interview and observation. Field data was generated using interview schedules. The schedules comprised of questions related to age-old conservation practices for conservation of natural resources in their habitats, issues and challenges faced by the Bonda women in preservation practices and suggestions or a future roadmap. Focus group discussions were also conducted in order to elicit detailed information from the Bonda women together as a group. Additionally, open ended interviews were conducted to allow the respondents to provide their views and opinions freely and without any specific limitations.

4.3 Contextualizing the Study

Women get disproportionately affected by climate change. The impacts of climate change intersect with existing gender inequalities, resulting in unique challenges for women in various aspects of their lives. The concept of intersectionality highlights how various forms of oppression and discrimination, such as those related to gender, race, class, and more, intersect and interact. This concept acknowledges that an individual's experience of oppression is shaped by the intersection of multiple identities (Samuels and Ross-Sheriff 2008). Both environmentalism and feminism recognize the importance of considering intersectionality to address complex issues effectively. Ecofeminism explicitly connects feminism and environmentalism. It suggests that the exploitation of women and the exploitation of the environment are interconnected issues rooted in patriarchal systems (Gebreyohannes and David 2022). Women, particularly in developing countries, often have a closer relationship with the environment due to their roles in agriculture and resource management. Environmental degradation can disproportionately affect women as they rely heavily on natural resources for their livelihoods (Balakrishnan 2023).

When assessing the significant effect of climate change on women in India, it is important to highlight a few of these interconnections in the context of the issues discussed in this chapter. In India, a significant portion of rural women's livelihoods is tied to agriculture. Climate change can disrupt agricultural patterns through erratic rainfall, prolonged droughts, and extreme weather events (USGCRP 2014). Women often lack access to resources such as land ownership, credit, and modern agricultural technologies, making it harder for them to adapt to changing conditions (Kristjanson Patricia et al. 2017). Women in India, like in many parts of the world, are responsible for fetching water for their households. As climate change leads to altered precipitation patterns and water scarcity, women often have to travel longer distances to collect water, further impacting their time and health (Yadav and Lal 2018). Rural women are primary users of household energy. Absence of clean and affordable energy sources affects their health and quality of life. Climate change exacerbates health issues, particularly for pregnant women and children. Heatwaves, changing disease patterns, and inadequate access to healthcare can disproportionately affect women's well-being (Cann 2013).

Women are generally more vulnerable during natural disasters due to socio-economic factors. Disasters like floods and cyclones can displace families and disrupt communities, putting women at risk of violence, trafficking, and exploitation (Nellemann et al. 2011). Climate-induced migration can lead to changes in family dynamics, with men often migrating for work leaving women behind to manage households. This can increase the workload and responsibilities for women, impacting their social and economic status (ibid. p. 20). Climate-related issues can impact girls' education. When resources are scarce, families tend to prioritize boys' education over girls', perpetuating gender disparities.

Despite these challenges, women often play critical roles in community resilience and adaptation strategies. Their knowledge of local ecosystems and traditional practices can contribute to sustainable solutions. Even though efforts are being made to address these challenges and integrate gender perspectives into climate change policies and initiatives, most often women are excluded from decision-making processes related to climate change adaptation and mitigation strategies (Smith et al. 2021). Their perspectives and experiences are crucial for creating effective policies.

Climate Change has a disproportionate impact on marginalized communities, especially women. Women are more vulnerable to the effects of climate change due to societal inequalities and their traditional roles (ibid. p. 10). Their reproductive rights also intersect with environmentalism, as overpopulation concerns sometimes clash with women's rights to make decisions about their bodies and reproductive choices (ibid. p. 22). Both environmentalism and feminism share common ground in their pursuit of justice, equality, and a better world. However, both these concepts have historically been criticized for lacking diversity and failing to consider the experiences of marginalized groups. Efforts to increase representation and inclusion are important in both spheres and this present study tries to fill this gap.

4.4 Tribes of Odisha and Bonda Tribe

The diverse cultural landscape of India is a source of fascination and wonder, encompassing a myriad of ethnic groups, languages, and traditions. Odisha is home to a significant population of indigenous or tribal communities, often referred to as Scheduled Tribes (STs) or Adivasis. These communities reside in hilly and forested areas, characterized by a varied cultural diversity, unique social structures, and a deep connection to the land. Every tribal community in Odisha is unique in their own way. They have a distinct language, traditions, customs, and artistic expressions of cultural identity that includes unique forms of music, dance, art, and oral traditions. Tribal societies generally have their own social hierarchies and systems of governance. The village council or assembly, often led by elder members of the community, plays a significant role in decision-making.

Many tribal communities practice subsistence agriculture, hunting, gathering, and fishing. Their livelihoods are often closely tied to the natural resources of their surroundings. They have traditional knowledge about sustainable resource management. Even their diverse religious beliefs often revolve around nature and spirits. Animism and ancestor worship are common elements of tribal religions. Many tribes have unique rituals and ceremonies associated with agricultural cycles, hunting, and other important life events.

Land is a critical aspect of tribal life. Many tribal communities have faced displacement due to development projects like dams, mines, and conservation efforts, leading to conflicts over land rights and cultural disruption. Tribal communities in Odisha often face social, economic, and political marginalization. They may lack access to quality education, healthcare, and other basic services. Poverty and lack of representation are significant challenges. There is an ongoing debate about development and its impact on tribal communities. Some argue for preserving traditional ways of life and protecting their rights, while others emphasize the need for education and modern amenities (ILO 2019). It is important to note that the situation and challenges faced by tribal communities can vary widely across different regions of Odisha. As the state continues to evolve, efforts to balance development with the preservation of tribal identities and rights remain ongoing.

The are many prominent tribes found in Odisha like Bonda, Kondh, Santal, Gond, Didayi, Juang, Munda, Ho, Koya and Paroja. This study particularly focuses on the Bonda tribal community residing in the remote hill regions in 32 isolated hilltop villages on the Kondakamberu mountain range of the Eastern Ghats in the Malkangiri district in Odisha. The Bonda tribe is one of Odisha's 13 Particularly Vulnerable Tribal Groups (PVTGs) that stands out as a unique and intriguing community. Nestled in the remote hills of the Eastern Ghats, primarily in the state of Odisha, the Bonda tribe offers a captivating glimpse into a way of life that remains rooted in tradition and history. They have been part of the primary influx of movement out of Africa several years ago. They were India's first forest settlers. The Bonda community had a population of 12,231 people according to 2011 census, with more women than men. Known for their distinctive cultural practices, the Bondas are a farming community that earns extra money by raising animals and harvesting seasonal forest products. Their culture is a tapestry of rituals, customs, and beliefs that has been passed down through generations. Their traditional attire is particularly distinctive, with both men and women wearing minimal clothing and intricate ornaments made from locally available materials. They are known for their distinctive attire, which includes colourful beadwork and headgear. The tribe is known for their intricate beadwork, which adorns their necks and wrists. The Bonda people have a unique social structure and economy. Malkangiri's geographical isolation has contributed to the preservation of the tribe's cultural practices, though it has also presented them with challenges in terms of connectivity and access to modern amenities. The Bonda tribe offers a window into a unique cultural mosaic that is an integral part of India's heritage. Their matriarchal social structure, distinctive attire, and strong ties to tradition reflect the beauty and diversity of human cultures. In recent times, extreme weather and climate change have had a negative impact on their patterns of earning a living.

4.5 Tribes and Environment—The Symbiotic Coexistence

The relationship between tribes and environmental protection is often complex and multifaceted. Many indigenous tribal communities around the world have deep cultural, spiritual, and traditional connections to their natural surroundings (Bruchac 2014). As a result, they often play a significant role in environmental protection and conservation efforts. In Odisha, the tribal communities possess a wealth of knowledge i.e., Traditional Ecological Knowledge (TEK) about their local ecosystems, including plant and animal species, weather patterns, and sustainable resource management. This knowledge, passed down through generations, contributes valuable insights to modern conservation practices. Many tribal societies have developed sustainable resource management practices that prioritize long-term environmental health. These practices are often rooted in their spiritual beliefs and cultural values, emphasizing the importance of living in harmony with nature.

The land, water, and natural resources hold deep cultural and spiritual significance for many tribal communities in Odisha. This connection fosters a strong desire to protect and preserve these resources for future generations. Some tribal groups like Gonds, Bondas etc. see themselves as stewards or guardians of their ancestral lands. This perspective leads to a sense of responsibility to care for and protect the environment from degradation and exploitation. In some cases, it is found that, securing land tenure and rights for indigenous and tribal communities can empower them to take a more active role in managing and protecting their lands and resources. When they have legal recognition and control over their territories, they are better equipped to implement sustainable practices.

Many tribal communities in Odisha engage in advocacy and activism to protect their land and resist activities that threaten the environment, such as deforestation, mining, and pollution. They often collaborate with non-governmental organizations, governments, and international bodies to raise awareness and seek support. Collaborations between indigenous groups, environmental organizations, and governmental agencies have led to successful conservation projects. These partnerships leverage the traditional knowledge and insights of tribal communities alongside modern conservation strategies.

The NGOs help in creating awareness among the tribals by demonstrating the conservation and preservation of the forest and its resources. Gradually, few young tribal local leaders have recognized the rights of indigenous people to manage and protect their ancestral lands. They are realizing that legal recognition and involvement in decision-making processes can enhance the ability of tribes to safeguard their environment. Many NGOs are promoting environmental education and awareness among tribal women with an aim to empower them to become active participants in environmental conservation and sustainability initiatives.

4.6 Underlying Gender Dynamics and Environment: Evidence from Field

It is essential to note that the relationship between tribes and environmental protection is not universally harmonious. Indigenous communities have often faced threats from activities driven by external interests like deforestation, mining and pollution. While conducting extensive field work in the tribal districts of Odisha, various issues related to gender and environment were traced. Displacement, cultural disruption, and loss of traditional practices are some of the challenges they encounter in their efforts to protect their environment. However, the impact of these troubled situations is more on the tribal women.

Environmental challenges such as climate change, deforestation, water scarcity, and pollution have different impacts on tribal women in Odisha, compared to men due to existing gender roles and inequalities. Women often have specific responsibilities related to water collection, food production, and energy use in many societies. Environmental degradation increases their workload and limits their access to resources, affecting their health, education, and well-being. Tribal women, in Odisha, are especially vulnerable to the impact of environmental disasters and climate change due to their poor socioeconomic status, constrained resource access and ownership and restricted decision-making power. During natural disasters, women may face challenges related to evacuation, healthcare, and protection.

In spite of these challenges, tribal women in Odisha have been at the forefront of environmental conservation and sustainability efforts. They often play a crucial role in natural resource management, biodiversity conservation, and sustainable agriculture. Recognizing and supporting their contributions can lead to more effective and equitable environmental solutions because there is evidence to suggest that empowering women can contribute to better environmental outcomes. When women have access to education, healthcare, and economic opportunities, they can make more informed decisions about family planning, resource management, and community development, which can positively impact the environment. The tribal women were found to be gradually on the path of empowerment.

Thus, advocacy for gender-inclusive environmental policies is essential. These policies should consider the specific needs and contributions of women, ensuring their active participation in decision-making processes related to environmental management and climate adaptation. However, the government is yet to make policies for the Bonda women and their forest rights. The tribal women are becoming aware that as the world transitions towards more sustainable practices, the emerging green economy could offer new opportunities for women in sectors like renewable energy, eco-tourism, and sustainable agriculture. However, they feel that it is important to harness this potential and look for sustainable future.

4.7 Bonda Tribes of Odisha

One of the most remarkable aspects of Bonda culture is their strong matriarchal social structure. Women play a pivotal role in decision-making and hold significant influence within their community. This unique arrangement contrasts with the predominantly patriarchal structure observed in many other parts of the country. The Bonda tribe primarily relies on subsistence agriculture for their livelihood. They cultivate crops like millets, pulses, and vegetables, using traditional farming methods. However, changing weather patterns and a shift towards cash crops have posed challenges to their agricultural practices. Collecting forest produce, hunting, and fishing is also integral to the tribe's economy.

Bonda tribes of Odisha, like many other indigenous communities, have gender-specific knowledge about their local environments. These women possess unique insights into natural resources and ecosystem dynamics. Recognizing and integrating this knowledge into broader environmental strategies can enhance sustainability efforts.

While the Bonda tribe's unique way of life is a testament to their resilience and adaptability, they also face numerous challenges in the modern era. Lack of proper healthcare facilities, education, and essential infrastructure continue to impact their quality of life. Additionally, their remote location has made it difficult for them to participate fully in the larger economic and social developments of the country. Their cultural identity also faces the risk of erosion due to external influences and urbanization. The younger generation, in particular, is exposed to external media and lifestyles, which could potentially dilute their traditional practices and values. Efforts are being made by various organizations, both governmental and non-governmental, to preserve and promote the cultural heritage of the Bonda tribe. Initiatives include providing better healthcare and education facilities, promoting sustainable agricultural practices, and creating platforms to showcase their traditional art and craftsmanship.

While the Bonda tribe faces numerous challenges in the modern world, their resilience and the efforts to preserve their cultural identity are testament to their enduring spirit. As we navigate the complexities of the twenty-first century, it is important that we respect and learn from the experiences of such indigenous communities to ensure the preservation of their rich heritage for generations to come.

Over a period in time, environmental change has significantly impacted their lives and livelihoods. Incessant rainfall washes away the ripe dirt from the slants. The oral history of Bonda clan describes how this whole local area relied upon '*Dangar Chas*' for survival and subsistence. '*Dangar Chas*' is a one-of-a-kind method of shifting cultivation that makes it possible to grow a wide range of crops without causing damage to the land. The *Chas*, according to the United Nations Framework Convention on Climate Change, is a highland crop that adds nutrition to the food supply without harming the environment.

4.8 About the Bonda Tribal Women

Central to the identity and vitality of the Bonda tribe are its women, who play pivotal roles in shaping the socio-economic fabric of their community. This section delves into the lives of Bonda tribe women, highlighting their empowerment, resilience, and contributions within the context of their traditional society. The tribe's matriarchal structure sets it apart from many other communities in India. Bonda women are at the heart of decision-making processes, and their opinions hold significant weight in matters of the tribe's welfare.

Bonda women have taken on vital roles within their society's economic sphere. Agriculture and livestock rearing are significant components of their livelihood, and women actively participate in these activities. They demonstrate remarkable agricultural expertise and are responsible for cultivating crops, such as millets and vegetables, which form the dietary staples of their community. Additionally, they manage the rearing of livestock, ensuring the availability of milk and other animal products for their families. Through these efforts, Bonda women contribute directly to the sustenance and survival of their community. Art and craftsmanship also hold a special place in Bonda society, with women showcasing their creative prowess through the crafting of intricate ornaments, pottery, and textiles. Their skills are not just a form of cultural expression but also a source of economic independence. The sale of these crafts provides a supplementary income for Bonda families, enhancing the financial stability of the household.

Bonda women have demonstrated a growing awareness of the importance of education for their children, particularly daughters. Despite their traditional lifestyle, they are increasingly realizing the value of education in breaking the cycle of poverty and ensuring a better future for their community. Many Bonda women are advocating for improved access to education and are actively engaged in encouraging their children, especially girls, to pursue schooling.

Bonda women play a crucial role in preserving the cultural heritage of their tribe. They are the carriers of traditions, passing down oral histories, songs, and rituals to the younger generations. Their roles in ceremonies and festivals are integral to maintaining the cultural identity of the Bonda tribe. The Bonda tribe women stand as an inspiring example of empowerment and resilience within the context of their traditional society. Their contributions to agriculture, craftsmanship, education, and cultural preservation underscore their importance in shaping the trajectory of their community's progress. As they continue to navigate the challenges of a changing world, Bonda women's determination serves as a reminder that cultural heritage and gender equality can coexist, creating a harmonious balance between tradition and progress.

4.9 Bonda Women, Indigenous Knowledge and Resilience to Climate Change

Malkangiri experiences a higher average annual rainfall of 1,667.6 mm than other districts in Odisha (https://malkangiri.nic.in/agriculture-change). But mono-crops are frequently destroyed by flash floods and landslides. Surface soil conservation has improved as traditional millet farming has been promoted, resulting in less siltation and erosion on the Bonda hills. The Bonda women are resolving the climate change issues by returning to the development of local millet assortments - farm (sanwa), finger (ragi), proso (chena), foxtail (kakum or kangni), and pearl (bajra) millets - which are climate-change resilient and guarantee the local area's food and wholesome security. The aforementioned indigenous climate-smart crops are providing these women farmers with numerous advantages.

Bonda, like other indigenous communities rely on subsistence farming, fishing, and hunting for their livelihoods. Climate change can disrupt these practices through altered weather patterns, shifting habitats, and changes in plant and animal behaviour. Bonda women are often more vulnerable to the impacts of climate change due to factors such as their reliance on traditional livelihoods, limited access to resources, and remote geographical locations. Bonda women often bear the responsibility of managing these resources and finding innovative solutions to sustain their families and communities.

The Bonda women possess traditional knowledge, a knowledge that refers to the skills and viewpoints that have been gathered through repeated encounters between communities and their natural surroundings. These agricultural methods and traditional knowledge have enabled them to adapt to changing climatic conditions, highlighting the significance of their practices in the context of climate change resilience. It is part of the local cultural traditions. These indigenous women are the keepers of cultural traditions, including knowledge about medicinal plants, storytelling, and rituals tied to the environment. Climate change can threaten these cultural practices, leading to a loss of identity and a weakened sense of community.

Bonda women play a pivotal role in building community resilience. They often manage local food systems, water sources, and healthcare. Their knowledge of sustainable practices helps the communities adapt to changing conditions and minimize negative impacts. These typical practices are discussed as follows:

1. *Mixed Cropping*: The Bonda women practice mixed cropping, where multiple crops are grown together in the same field. The practice of mixed cropping minimizes the risk of complete crop failure during extreme weather events, as different crops have varying resilience to climate stressors. This approach minimizes the risk of crop failure due to erratic weather patterns and provides a diversified source of food and income. Examples of crops grown include millets, pulses, tubers, and vegetables.
2. *Agroforestry*: The integration of trees and shrubs in their agricultural fields helps regulate temperature, reduce soil erosion, and improve water retention. These trees also offer additional resources such as fruits, firewood, and medicinal plants.
3. *Indigenous Seed Varieties*: Bonda women have preserved and continue to use indigenous seed varieties adapted to their local climate. These seeds are naturally suited to the region's changing weather patterns, ensuring better crop yields and resilience. Bonda women watch to it that the seeds are organically preserved and sun-dried. They safeguard the seeds by covering them with bengunia and neem leaves. This is a characteristic method of seed conservation inferable from the customary information among Bonda women.
4. *Water Management*: They have developed intricate rainwater harvesting and irrigation systems to deal with erratic rainfall patterns. Traditional water management practices like dugout ponds and small dams help store rainwater for dry periods, ensuring a consistent water supply for crops.
5. *Minimal Dependence on Modern Inputs*: They rely minimally on chemical fertilizers and pesticides, opting for organic and sustainable farming practices. They do not use pesticide and instead encourage birds and other animals to eat pests that could ruin crops. This reduces their vulnerability to market fluctuations and economic challenges associated with climate change.
6. *Traditional Knowledge*: The Bonda women's deep understanding of local climate patterns, based on generations of observation, allows them to time their planting and harvesting to maximize crop yields even under changing weather conditions. They possess a deep understanding of their local ecosystems, including weather patterns, plant and animal behaviour, and natural resource management. This traditional knowledge is invaluable for adapting to changing climatic conditions and being resilient to it.

The Bonda tribal women's agriculture serves as a valuable example of how indigenous knowledge and traditional practices can contribute to climate change resilience. Their sustainable farming methods, deeply rooted in their cultural heritage, offer insights and lessons for modern agricultural practices, highlighting the importance of preserving traditional wisdom in the face of a changing climate. Despite facing systemic barriers, women from the Bonda community are emerging as leaders and advocates in climate change discussions. They bring a unique perspective and can amplify the voices of their communities on national and international platforms. It is essential to recognize and support such communities as they play a pivotal role in the global efforts to combat climate change and promote sustainable agriculture. These women play a significant role in addressing and coping with the impacts of climate change due to their close connection with the environment and their traditional knowledge.

It's important to recognize that gender dynamics within Bonda communities can influence how women and men experience and respond to climate change. Women might have different roles and responsibilities, and their contributions are sometimes undervalued or marginalized in decision-making processes. The Bonda women opined that efforts to address climate change must be inclusive and consider the specific needs and contributions of women. These initiatives include involving them in policy discussions, resource allocation, and adaptation planning.

Land rights are often crucial for indigenous communities, as they are closely tied to their cultural, spiritual, and economic well-being. Bonda women are central to land stewardship, because they have stayed with close proximity to nature and have been conserving the natural resources over generations. The Bonda women have taken up collective responsibility to carefully manage the land and ensure the quality and abundance of the natural resources, especially comprising of forest land and securing their land rights is essential for their climate change adaptability.

4.10 Conclusion

Historically, tribal women have had a significant impact on the preservation of their traditional heritage, especially responsible resource management. They have grown and provided food for their local communities. They also have been the guardians of agricultural technology maintaining the multi-cultural biodiversity conservation methods (International Indigenous Women's Forum Declaration 2005). Nonetheless, the tribal communities, academia and the civil society stakeholders agree upon the fact that there has been a dwindling impact on women's traditional roles (GOI 2015). This has led to a large number of tribal women living on the periphery of society in India. They experience numerous forms of prejudice as women and as indigenous people. They experience severe poverty, human trafficking, illiteracy, negligible lands rights and substandard health care services, and both private and official violence (GLRF & CWLR 2006). Women find it relatively problematic for overcoming the ill-impacts of natural disasters and depleting biodiversity on their livelihoods and lives. These results due to the complicated web of severe constraints that worsen their poverty.

Nevertheless, the case of the Bonda women in Odisha exemplifies a best practice which can be replicated elsewhere. However, such stories go undocumented and therefore, remain in oblivion. The resilience exemplified by the Bonda Women is a testimony to the fact that tribal women are closer to nature, and they would put in their best efforts to protect it. Their inert passion to thrive against all odds is an exceptional case. However, many tribal women across India succumb to the rapid and faulty development agenda that catalyses climate change. They remain excluded from the development euphoria.

Therefore, it is important to strengthen the tribal women participation and include the gender experts in any bureaucratic policy level planning and decision-making exercise concerning climate change. The consultation and decision-making processes should be women-inclusive and participatory as they have a direct impact on their livelihoods. In summary, the unique knowledge, roles, and responsibilities of indigenous women make them vital actors in addressing the challenges posed by climate change. While the Bonda tribal women's agricultural practices have demonstrated remarkable resilience to climate change, they are not immune to the broader challenges of environmental degradation, land encroachment, and changing societal dynamics. To ensure the preservation and continued success of their sustainable farming methods, it is crucial to provide support in the form of sustainable development initiatives, education, and access to resources.

The Indian government and various organizations are recognizing the importance of empowering women and promoting gender equality as part of broader climate action strategies. However, the question that remains unanswered is the context of tribal women in India and addressing their issues and vulnerabilities. Tribal women in India face unique challenges related to climate change due to existing gender inequalities and socio-economic factors. Addressing these challenges requires a holistic approach that considers the roles and contributions of women in both climate impacts and adaptation strategies. Recognizing their contributions, respecting their traditional knowledge, and including them in decision-making processes are essential steps toward building resilience and sustainability in the face of Climate Change and Environmental hazards.

References

Agrawala S, Fankhauser S (2008) Economics aspects of adaptation to climate change. Costs, benefits and policy instrument. OECD, Paris

Balakrishnan S (2023) Empowering the unheard: Why women's voices are crucial in environmental policy and action. International Union for Conservation of Nature and Natural Resources

Bruchac MM (2014) Indigenous knowledge and traditional knowledge. In: Smith C (eds) Encyclopedia of global archaeology. Springer, New York, NY. https://doi.org/10.1007/978-1-4419-0465-2_10

Cann KF (2013) Extreme water-related weather events and waterborne disease. Epidemiology and Infection. https://pubmed.ncbi.nlm.nih.gov/22877498/

Gebreyohannes NM, David AD (2022) Women and nature: an ecofeminist reading of Chimamanda Ngozi Adichie's purple hibiscus. Literature 2(3):179–188. https://doi.org/10.3390/literature2030015

GLRF & CWLR (2006) We know what we need, South & consult for women and land rights-establishing women's resource Asian Women speak out on climate change rights agenda in tribal community. Ranchi

GoI (2015) Report of the high level committee on the status of women in India, Ministry of Women and Child Development, vol 4
IAASTD (2009) Agriculture at a crossroads. The Island Press, Washington DC
ILO (2016) The labour situation of indigenous women in Peru: a study. Geneva, ILO
ILO (2019) Indigenous & tribal peoples' rights in practice, programme to promote ILO Convention No. 169 (PRO 169) International Labour Standards Department, 2009
Kristjanson Patricia B, Elizabeth QB, Twyman J, Meinzen-Dick R, Kieran C, Ringler C, Jost C, Doss C (2017) Addressing gender in agricultural research for development in the face of a changing climate: where are we and where should we be going? Int J Agric Sustain 15(5):482–500. https://doi.org/10.1080/14735903.2017.1336411
Nellemann C, Verma R, Hislop L (eds) (2011) Women at the frontline of climate change: gender risks and hopes. A Rapid Response Assessment. United Nations Environment Programme, GRID-Arendal. ISBN: 978-82-7701-099-1 Printed by Birkeland Trykkeri AS, Norway
Samuels G, Ross-Sheriff F (2008) Identity, oppression, and power: feminisms and intersectionality theory. Affilia-J Women Soc Work 23:5–9. https://doi.org/10.1177/0886109907310475
Smith JM, Lauren O, Jennifer Grosman F (2021) The climate-gender conflict nexus amplifying women's contributions at the grassroots. Georgetown Institute for Women, Peace and Security
UNESCO intangible cultural heritage, individual case studies 1992–2023
UNISDR (2006) Words into action: implementing the hyogo framework for action document for consultation draft, November
United States Global Change Research Programme (USGCRP) (2014) Hatfield J, Takle G, Grotjahn R, Holden P, Izaurralde RC, Mader T, Marshall E, Liverman D. Ch. 6: Agriculture. Climate change impacts in the United States: The Third National Climate Assessment, Melillo JM, Terese (T.C.) Richmond, Yohe GW (eds). U.S. Global Change Research Program, pp 150–174
Yadav SS, Lal R (2018) Vulnerability of women to climate change in arid and semi-arid regions: the case of India and South Asia. J Arid Environ

Chapter 5
Reducing the Risks of Transboundary Climate Change Impacts in India and Bangladesh: Options for Cooperation

Nisha Thankappan

Abstract Risks and vulnerabilities of climate change impacts are increasing regardless of adaptation and mitigation efforts. Borderless climate change has cascading effects that strain transboundary resources and countries' response capacity. Transborder climate risks are climate change-triggered events that originate at a specific location, and the area of origin shares its impacts with its immediate or distant neighbouring countries or regions via multiple pathways. Bangladesh and India are confronting cross-border climate change impacts on shared natural resources like Trans- Himalayan River water resources and Sundarbans Mangrove Forest. The occurrence of natural disasters in Bangladesh leads to cross-border migration to India, which blurs their territorial boundaries and incites tensions between them. This paper argues that adaptation without borders is mandatory for effective risk reduction. This study contextualises cooperation and coordination of shared resources management and adaptation beyond a single government's effort at the national level. This study evaluates how transboundary cooperation in adaptation could solve climate risk emergencies and avoid conflicts between India and Bangladesh. For this, the study used the method of content analysis and narrative review that drew from the literature to understand the underlying concepts and problems connected to cross-border climate change, international cooperation, and adaptation.

Keywords Climate change · Cross-border risks · Transboundary adaptation · Cooperation · India · Bangladesh

---

N. Thankappan (✉)
Centre for South Asian Studies, School of International Studies, Jawaharlal Nehru University, New Delhi, India
e-mail: nisha.s@lamperofos.com

Climate Change Research Division, Lampero Fos Research Consultants Pvt. Ltd, New Delhi, India

A. Sarkar et al. (eds.), *Risk, Uncertainty and Maladaptation to Climate Change*, Disaster Risk Reduction, https://doi.org/10.1007/978-981-99-9474-8_5

5.1 Introduction

Border-less adaptation to climate change is difficult yet essential. Globalisation has brought the idea of borderless engagement of people-to-people contact, the trans flow of finance, products, and resources and now transnational climate change impacts. Emerging climate change risks prove that the international community must reshuffle the processes and channels through which adaptation occurs. The existing and agreed adaptation frameworks are ardent towards looking at the issue within the spectrum of local, national (Lesnikowski et al. 2016) and international contexts. The current adaptation mechanism corresponding to the direct effects of climate change within a territorial boundary is governed by few local or national authorities (Banda 2018). Most local or national government policies are deliberately responsive rather than proactive (Challinor et al. 2018). Until recently, "it is rare to identify cases of regional cooperation to *jointly manage* shared climate risks that cross borders" (Benzie and Persson 2019).

In the recent period, "there is growing recognition that adaptation also must consider 'internal' changes, such as those occurring within geographical regions as a consequence of the adaptation practices of others. Adaptation is conditioned by capacity, which includes adequate funding, trained personnel, and access to and an ability to use relevant information" (Scott et al. 2012). Thus, an increasing trend in expanding the climate risk within the territorial framework is being challenged and demanding a transnational understanding. Several reasons are in line to substantiate this move concerning climate change's multilateral impacts and implications. Studies demonstrate that transnational climate risks seriously hit supply chains, security, and finance flow, and trigger unintended people flow. Subsequently, this realisation has challenged the idea of administering adaptation within the local and national territory framework. The significant reasons are two-fold. Firstly, the fundamentals of adaptation were registered within the climate regime by mapping out adaptation as a global-scale challenge precisely in the Paris Agreement 2016. Secondly, from outside the climate regime, owing to the growing concerns of the security institutions that are tying climate change risks and global risk management together in a general framework (Benzie and Persson 2019).

Adaptation and risk management of transboundary resources in South Asia is a matter of concern for the development and security of the region. Bangladesh is the classic example of proneness to natural disasters and climate vulnerability in the world, while India is one of the major climate change hotspots in South Asia. India and Bangladesh suffer most from Tropical Cyclones (Ali 1999). They are confronting climate change-triggered repercussions on human mobility and shared natural resources like trans-Himalayan water resources and Sundarbans Mangrove Forest. Apart from this long-standing dispute and conflict between them regarding water sharing, it is a classic example (Das and Bandyopadhyay 2015) that will be exacerbated by future climate change. Against this backdrop, the key objective of this paper is to examine transboundary rivers between these countries and how they impacted by climate risk and adaptation.

Climate change also triggers natural disasters and transborder people migration. This leads us to question whether territorial flexibility and governance are sufficient to manage transboundary risks and hazards. This chapter tries to answer questions like; What does borderless governance entail, and what are some potential response methods? Can India and Bangladesh collaborate on a bilateral and regional basis to adapt to transnational climate risks?

To best answer the research objectives, the paper uses a case study methodology and an interpretive approach to data analysis. The information was gathered by conducting content analyses on several published policy documents, reports, and press releases of the Intergovernmental organisations governments of Bangladesh and India. These include ministerial-level meeting press releases of governments of India and Bangladesh on Water Sharing, the IPCC Synthesis Reports, and the Asian Development Bank Report on Climate change Adaptation in South Asia, Content analysis was chosen to investigate the concepts, narratives and policy results portrayed in the literature.

5.2 Borderless Climate Risk and Potential Significance

A well-grounded or universally accepted definition of cross-border climate change impacts is yet to be born. The Fifth Assessment Report of the Inter-Governmental Panel on Climate Change (IPCC 2014) has briefly touched upon the trends of cross-regional or transboundary implications of climate change. This report recognised this phenomenon as an '*indirect*' or '*transboundary*' or '*long-distance*' impact of climate change. It further stated that the global trading systems are threatened by climate change impacts, especially on food trade. Food trade between the countries will be most at risk because of reduced food productivity, subsequent price escalations, and affordability during extreme weather changes in a country (IPCC 2014). Further to this, Ercin et al. (2021) reports that the European Union's (EU) agri-food economy will be under moderate vulnerability due to drought in non-EU countries in the future. However, the IPCC report has no thematic visionary content on this propounded risk.

Several studies have linked the cross-border climate risk with the adaptation framework. The "*nested concept of vulnerability*" approach portrayed vulnerability and adaptation as a 'tele-connected' phenomenon that emphasises on any transmission of a coherent effect beyond the place where the initial force of action originates. Teleconnection defines the correlation of events and highlights the necessity to identify connecting mechanisms, drivers, and net outcomes. It has been argued that "the concept of tele-connected and nested vulnerability is fundamental as countries move forward to reduce vulnerability to the anticipated impacts of climate change while recognising that social processes are as complex and dynamic as the climate system" (Eakin et al. 2008).

The initial works of the conceptual narration of the broader perspective of cross-border impacts of climate change initiated by the Stockholm Environmental Institute have used the term "indirect climate impacts" (Benzie 2014). Pescaroli and Alexander

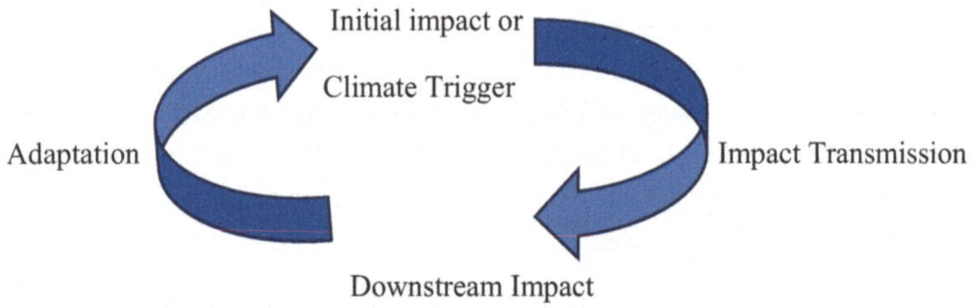

Fig. 5.1 Cross-border climate impact transmission. *Source* Author's own elaboration

define "cascade effects implying the initial effects can trigger other phenomena that lead to consequences with significant latitudes. These effects are complex and multi-dimensional and evolve constantly over time. These phenomena are thus primary, secondary, tertiary and so on" (Pescaroli and Alexander, 2015). Similarly, Goldain used the term "Cascading Risk" to propagate the emerging challenges of the globalised era, climate change, migration, cyber security, etc. (Goldin and Mariathasan 2014). Several other terminologies got attention in a short span of time, like "Connected Risk" (Galaz et al. 2014), or "double exposure" (Leichenko et al. 2010). A detailed approach to developing this framework started within the ecology literature (Fig. 5.1).

A direct climate change impact in one nation may initiate an indirect impact in another nation or region. Benzie et al. (2016) defined transnational climate change as "direct impacts in one country may be transferred by various flows to affect another country". These impacts are identified under four significant pathways: bio-physical, finance, trade, and people for the effective formation of adaptation techniques (Benzie et al. 2016).

Transboundary climate change risk has to be governed beyond the territorial framework set by the epistemic community, which is rendered in the powerful base of environmental science. The new momentum of the governance structure is propounded at three levels—national and bilateral, transnational, international and regional—with the International Relations community's systematic evaluation (Benzie and Persson 2019).

Further, Carter et al. (2021) defines cross-border climate change impacts as "consequences of climate change that occur remotely from the location of their initial impact, where both impacts and potentially also responses to those impacts, such as adaptation, are transmitted across one or more borders" (Carter et al. 2021). It refers to two major components: climate change impacts and adaptation. After that, the authors (Carter et al. 2021) identified and detailed seven pathways of cross-border climate change impacts as follows:

- *Trade*: It implies the flows of commodities, goods, and services on international markets
- *Finance*: Change in the value and flow of public and private capital
- *People*: The large flow of people across borders as migrants and movement of people such as tourism

- *Psychological*: A cognitive filter arises out of the external sources about the risk and consequences associated with the climate change impacts through media and other actors
- *Geopolitical*: The political phase of the impact arises in international affairs, resource allocation, strategy, and adaptive measures
- *Bio*-Physical: Hydrological systems and flow of water, movement of pests, species, and pathogens
- *Infrastructure*: Impact upon the telecommunication networks, transport accessibility

This chapter analyses cross-border climate change impacts between India and Bangladesh in three scenarios: bio-physical, geopolitical and people migration under the people's pathways propounded by Carter et al. (2021).

5.3 Adaptation and Vulnerability

The communities and households in the ecosystem they belong to are susceptible to exposure and sensitivities to various ecological, economic, and social changes. Also, they have different or similar kind of capacity to anticipate, adjust and cope with these changes. Intergovernmental Panel on Climate Change (IPCC) defines adaptation as "…adjustment in natural or human systems to a new or changing environment. Adaptation to climate change refers to adjustment in natural or human systems in response to actual or expected climatic stimuli or their effects, which moderates harm or exploits beneficial opportunities. Diverse types of adaptation can be distinguished, including anticipatory and reactive adaptation, private and public adaptation, and autonomous and planned adaptation" (IPCC 2001).

Vulnerability defines the communities' function of exposure to a system, sensitivity to multiple shocks and stress associated with the environment, economy or society, and their variability in capacity to adapt to such adverse changes. Adger (2009) argue that vulnerabilities in socio-ecological systems are interdependent and connected across different localities and peoples. Also, adaptation measures taken for the safety, security, and welfare to an adverse economic or environmental change in one region may cause socio-ecological vulnerability in a different region or society and vice versa directly or indirectly. Therefore, the vulnerabilities are interdependent through various mechanisms and channels through which exposure and sensitivity are transmitted and the processes that affect adaptation capacity. The world is interconnected through mechanisms like global environmental change, the changing structure of the market economy, and material flows of people, resources, and information. Given this, "environmental change in one locality is increasingly connected to regional and global systems through human actions and responses" (Adger 2009). Therefore, cooperation and adaptation should be coordinated by mapping inter-country/inter-regional dependencies.

5.4 Risks and Cooperation on Transboundary River

The Tibetan Plateau, "*Third Pole*," in the Himalayas, is the source of origin of river bodies primarily running through Central, South, and Southeast East Asian regions. These regions' high priority on bilateral and multilateral water management is concentrated on the Ganges–Brahmaputra-Meghna System, which spans across Bangladesh, Bhutan, India, Nepal, and China. The Ganges River basin occupies the largest population of around 655 million in an area of around 1.2 million km. Agricultural livelihood activities in the area withdraw the highest amount of water from it. Also, its water is being contaminated with chemical composites, shrinking the river's health. The emerging water security crisis cannot go unnoticed by the riparian governments.

Transboundary water resources in the South Asian region are critical concerning climate change's transnational impacts. It is associated with the region's water availability, supply, and hydrological matrix, which depend solely on glacial melting and rainfall (Prabhakar et al. 2018). The foremost reason could be attributed to Bangladesh's sole dependence on perennial rivers of the unique transboundary water reservoir of Himalayan Glaciers such as the Ganges (known as 'Padma' in Bangladesh) and Brahmaputra (ICIMOD 2009). From the Ganges and others, Bangladesh receives over 91% of its water from India and a sizable share flow from Tibet (Chellaney 2014). Ganges traverses around 2,510 km through India and Bangladesh (Pandey 2014). The water flow in the upper catchment of Ganga and Brahmaputra is a combination of both rainfall-runoff having a 16% contribution while glacier melt contributes 11%, which is the primary life-sustaining source of Bangladesh for surface water.

The water resources shared between India and Bangladesh are a matter of high priority in the era of global warming. The shared area, particularly the rim of the Ganges, the Brahmaputra, and the Meghna rivers, known as the Ganga–Brahmaputra-Meghna Basin (GBM Basin) between Bangladesh and India, is strategic and geomorphologically pivotal for both countries. The Ganges River has often been disputed among the transboundary rivers shared between India and Bangladesh, with disagreements on various levels. Out of these, the deliberation of water sharing, and conflictual environment is concentrated precisely on the India-built Farakka Barrage in the Ganges that runs 15-mile upstream on the border of Bangladesh. Until now, the regulation of Ganga water in Bangladesh is governed under the Indo-Bangladesh Treaty of 1996 (GOI 2022a, b; Saklani et al. 2020). No significant development has been made so far.

In the era of climate extremes, the average standard or peripheral management of transboundary water is insufficient to meet the lives and livelihood of the people. Water scarcity for agricultural activities in the changing environment threatens farmers with reduced yield, significant loss of income and employment, and migration to urban areas. A study among wheat farmers in Iran's Meku city revealed that the lack of access to sufficient water resources for drinking and agricultural production has forced them to migrate to urban centres from rural areas. They would be willing to return to their respective villages if the water problem gets resolved and well-managed properly (Salimzadeh and Katantari 2015).

Similarly, the northwestern region of Bangladesh is critical to prolonged drought hours and water scarcity. Tangon River basin, an area of about 2388.88 km^2 in the Barind tract, stretches 267 km long between India and Bangladesh. As a transboundary river, Tangon is a source of fresh water for the people of West Bengal and Bihar in India. More precisely, its downstream tributary, the Mahananda River in Bangladesh, has found a shortage of 63% of its flow volume due to the damming effect. Farmers and fishermen who depend on this river for their livelihoods are affected by water scarcity for their daily needs. This river was used to divert water from the Tesesta River till 1787 before the neo-tectonic upliftment of this alluvial plain. The eastward shifting of the Teesta River has transformed the Tangon into a non-perennial river, fed by only rainwater. It has severe implications for the groundwater recharge and storage in the lower riparian zone. Eighty percent of its total annual rainfall is received during monsoon, and the latter seasons face acute water scarcity (Pal et al. 2020). A study predicts that if the situation continues like this, Bangladesh will be another Yeman in the near future (Miyan 2015).

Transboundary water resources in South Asia, particularly between India and Bangladesh, are interlinked with geopolitics, demography, asymmetric power relations, and socio-cultural and political constraints (Biswas 2011). One of the significant constraints in the transboundary rivers' cooperative governance is the grouping of political leaders, bureaucrats, engineers, and other stakeholders. These groups disregard the opinions of the local population about their perspectives on lives and livelihood when it comes to river water decision-making. The growing trend, like water sharing, will become more complicated due to the gap between water availability and high demand. In this scenario, climate change and global warming play an active role in further deteriorating the situation, which can affect the sovereignty and territorial integrity of the countries. In the present era, the links between water security, and climate change are complex to determine its multi-dimensional impacts and no country can act independently or have arbitrary sovereignty over shared water resources.

Therefore, managing shared water resources from the adaptive perspective requires high-level dedication and shared interest between the riparian regions. The inclusion of science policy into the management would need "inclusivity, involvement, interaction, and influence" (Scott et al. 2012). Concerning the South Asian Himalayan water system, China is a consideration. To address the water situation in the northeast, China is using dam construction to move water from the upstream Brahmaputra. As a result, the lower riparian regions in Bangladesh and India are experiencing severe water shortages and are in a difficult position (Sullivan 2011). The diversion of water for irrigation and hydropower generation by the countries creates a conflict-driven situation among the riparian countries. It must be controlled through bilateral dialogues and cooperation. In the case of Bangladesh and India, the nature of political divisiveness between the Indian provincial states and the centre also contributes to the issues. If it could be resolved, significant cooperation can be done in managing transboundary water resources jointly, inland navigation and water transit, multipurpose storage dam projects, and joint management of the Sundarbans can be the potential areas for benefit sharing in the Ganges basin.

Such transmitted climate change impacts beyond borders can be resolved only through regional cooperation (Benzie et al. 2016). The "mutual suspicions and reluctance to cooperate between riparian may impair timely approaches to the collective action problems of non-traditional security threats such as water conflict" (Hanasz 2014). However, cooperation and agreements alone do not necessarily ensure equal water allocation. Such transboundary water security concerns have existed in other countries for centuries like the contestation on the natural water sharing of the Jordan River Basin that resulted in a buffer zone of conflict, and Nile water diversion between the Middle East and Africa (Mirumachi 2013).

5.5 Sundarbans: Conflict, Cooperation and Adaptation

The UNESCO World Heritage site of Sundarbans shared between Bangladesh and India is a unique biodiversity of the world's largest mangrove forest. Sundarbans consist of evergreen natural forests with rich biodiversity and habitats. It is situated in the lower deltaic estuaries of the GBM basin at the mouth of the Bay of Bengal. It spreads across 6,017 km^2 in southwest Bangladesh, constituting 62% of its area, and extends 4,246 km^2 in the southeastern part of West Bengal in India, constituting 38% of its land area. It occasionally receives freshwater runoff and sediments from the GBM distributaries (Ahmed et al. 1999). Being a buffer transition zone between land and sea, this forest is salt tolerant and produces organic sustenance and nutrients. However, this rich ecosystem is reeling under global climate change due to anthropogenic pressure. The rising sea level affects the freshwater inflows in the Sundarbans region along with recurring tides, cyclones, and associated floods on its marshy lands. Shared leadership and transboundary cooperation are required to conserve and protect this fragile area.

Several non-binding agreements have been made between Bangladesh and India, focusing on their shared natural resources, including the Sundarbans. A bilateral agreement signed on September 6, 2011, agreed upon enhancing bilateral cooperation for mutual benefits in a wide range of areas under the banner of trade, connectivity, and environment. These contained (a) promotion of water resources, (b) management of natural disasters, (c) environmental protection and responding to challenges of climate change through adaptation, (d) sub-regional cooperation in water resource management, environment and sustainable development and enhancing security cooperation (GoI 2011).

The optimum utilisation of the benefits of their shared understanding and vision through mutual collaboration about the environment and climate change is appreciable. A joint cooperation mechanism was proposed to guard the development and promote the conservation and resilience of the vulnerable landscape. Nevertheless, these efforts have yet to reach a formal agreement. However, a joint platform for the knowledge sharing and conservation of Sundarbans bio-diversity and flora and fauna began with the BISRCI (Bangladesh-India Sundarbans Region Cooperation Initiative). Under this, a Joint Working Group (JWG) on Conservation of Sundarbans (SAWI 2019) was constituted.

Despite this, the MoU between India and Bangladesh failed to effectively manage sea level rise, coastal erosion and flooding, loss of land, and deadly cyclonic storms in the Sundarbans area. The mandate of the MoU would focus on bilateral cooperation in the conservation of Sundarbans to tackle endangerment and extinction. This would ensure Sundarbans Mangrove Forest's capacity to serve as a vital protective mechanism against flooding, tidal waves, and cyclones. Secondly, the cooperation must work for the sustainable exploitation of natural resources for poverty alleviation and livelihood support rather than encroachment.

5.6 Trans-Himalayan Climate Collaboration

The Himalayan Mountain range is the high peak that separates Tibetan Plateau from the Indian sub-continent. People in the Indian subcontinent depend heavily on Himalayan resources for their needs, especially for water, hydropower, fisheries, transport, timber, etc. The recent trend in glacial melting, including Glacial Lake Outburst Floods (GLOF), cloud bursting, and natural disasters, is increasing tremendously in the Himalayas (Das and Bandyopadhyay 2015). These changes exponentially affect the Himalayan River basins as glaciers play an essential role in the hydrological cycle (Pramanik and Bhaduri 2016). In particular, the Ganges is predicted to experience higher runoff, the Brahmaputra to have higher flooding, and the Indus to be effect by large volume of melted water (Nepal and Shrestha 2015). At this point, cooperation on Himalayan River water management amongst the neighbouring nations becomes crucial. Substantial warming in the Hindu Kush Himalaya (HKH), where the most significant and extensive permanent ice cover is located outside the North and the South poles, exhibits reduced snowfall and glacier retreat. The projected trends of the increase of temperature in the HKH region for the near future (2040–2069) is 2.2 ± 0.9 °C (3.3 ± 1.4 °C) under the RCP4.5 scenario, whereas under the extreme scenario of RCP8.5, the temperature change projected to be 2.8 ± 1.2 °C (4.8 ± 1.7 °C) for both future scales of the twenty-first century (Sabin et al. 2020). Thus, expert intervention in glacial research is highly required to control unpredictable glacial lake flooding. Given trans border cooperation for sustainable mountain development, especially in the Hindu-Kush Himalayan region, several initiatives are in the picture, such as 'The Kailash Sacred Landscape Conservation and Development Initiative' (KSLCDI) (Anisimova and Magnan 2023).

5.7 Risk-Informed Governance and Management

Transboundary risk management is crucial for adapting to the damaging consequences of climatic influences across boundaries. It is necessary for the states to collaborate on specific grounds to coordinate and manage themselves thoroughly and transparently. In some regions, cooperation and partnership also hinder the sustainability of the environment. One of the example is that of the Rampal Power plant which is a joint venture of India-Bangladesh energy cooperation. The Bangladesh India Friendship Power Company (BIFPC), a shareholding company of Bangladesh's state-owned Bangladesh Power Development Board (BPDB) and India's National state-owned National Thermal Power Corporation (NTPC), collaborated for the Rampal coal-fired project in 2017. It has been proposed to be built at the UNESCO World Heritage site Sundarbans to meet the challenges of energy poverty of Bangladesh covering an area of over 1,834 acres of land in the Southwest of Bangladesh. The Rampal Power Project is supposed to be the largest power plant in Bangladesh with an estimated potential to produce 1320 MW power.

The disparity in political and economic benefits of this project for both India and Bangladesh are under question. The construction of infrastructural and economic developments on the bank of the Hugli River in West Bengal and its proximity to the Sundarbans area complicates the sustainability of this unique ecosystem. Concerns were raised about the environmental security implications of the plant in its operational phase since the plant is being constructed just 14 km from the reserved Sundarbans Forest. The Rampal Thermal Power development project has faced a massive setback from the opposition created by local communities and environmentalists. Increasing deforestation, glacial melting, reduced salinity in the upstream river flow, and overflow impinge upon the fisheries. For the local communities, this project is a threat to their livelihood and culture. People opine that this portrays a lack of political will of Bangladesh's government that has enormous negative implications on its people and its vulnerable environment. Moreover, agricultural land acquisition and conversion by the state and stakeholders for this plant has brought an environmental conflict situation in Bangladesh. The Bangladesh Poribesh Andolon Bangladesh (BAPA) are at the forefront of the protests. They argued that the power plant would release pollutant chemicals dangerous to the Sundarban's habitat (Islam and Al-Amin 2019). It would bring serious environmental sustainability challenges as it is implemented in the world's largest mangrove forest (Huda 2020). Therefore, the ecological feasibility of the power project is yet to be reconciled with the local people who depend on Sundarbans.

5.8 Natural Disasters and Displacement

The security paradigm of the state-centric approach has been narrowed with the overarching emphasis on non-traditional security threats like climate change. However, environmental security is considered a soft issue in the joint security agenda of the South Asia region. The states must identify that the security of a state is contingent on others, especially its neighbours since these countries not only share geographical boundaries but also share natural resources and have socio-demographic continuity. In this context, there is bound to be many unavoidable security challenges that are arising between India and Bangladesh due to climate change induced disasters.

Bangladesh is a country that experiences the worst impact of flooding particularly by riverine floods that happens annually. Interestingly, the highest flood-affected areas of Bangladesh often do not experience heavy rainfall. The floods in Bangladesh can be categorised into three types: flash floods, fluvial/tidal floods (Haque et al. 2018) and transboundary flow-induced floods or riverine floods. Flash floods in the northeastern region of Bangladesh are always triggered by the rainfall received in the adjacent northern part of the Meghalaya Plateau in India. Conversely, riverine flood results from high-intensity precipitation within the GBM catchment. Along this site, the Himalayas Mountain region of Bangladesh, however, plays a vital role in the floods in the downstream areas of Bangladesh. The heavy rainfall in the Himalayas causes a landslide that produces debris and sediment in the river channels, causing excess water accumulation. Such accumulated water overtops the barrier and flows down in a considerable volume. This type of water discharge has caused devastating floods in the downstream areas. Bangladesh experienced such a river flood in its history in 1968 in Jalpaiguri due to the breaching of temporary dams in the bed of the Teesta River (Rudra 2018).

Flood forecast and risk control are significant transboundary management issues in South Asia, especially between Bangladesh and India. A significant development in the river management on demands of the people has been realised in the 38th meeting of the ministerial-level Joint Rivers Commission of India and Bangladesh held in New Delhi in 2022. This joint meeting recognised the bilateral issues of river water sharing of common rivers. The highlight of the MoU between both countries for cooperation is that India will assist Bangladesh by sharing real-time flood data to help Bangladesh address unforeseen flood events. Out of the fifty-four rivers, of which 7 were identified earlier on a priority basis for developing a framework of water sharing agreements (GoI 2022a, b). Other aspects of this MoU addresses river pollution, conducting joint studies on sedimentation management, riverbank protection works, and water sharing of Kushiyara River (GoI 2022a, b). The Kushiyara River agreement is the first treaty since the Ganga Water Sharing Treaty of 1966. However, the contestation on Teesta River remains unresolved.

Despite these agreements, actions are terribly dead on the ground. The joint River Commission of India and Bangladesh was established in 1972 in view of the water management and sharing between these countries. However, a remarkable attention on the hydrological disasters, which are transboundary in nature, started only recently.

Bangladesh lacks the technical expertise and resources to mitigate the climate-posed threats successfully. Thus, India has extended augmenting Bangladesh's Humanitarian Aid and Disaster Relief Capabilities and is jointly working towards achieving energy security through green energy projects. Community-level cooperation in the water sector is an encouraging opportunity for India and Bangladesh. This can include an early warning system and flood information to manage climate risks in the bordering villages. This could save lives and property from the sudden onset of events. For instance, this mechanism was effectively introduced in Nepal and the Bihar State in India, where the Koshi River flood occurs annually (Molden et al. 2017).

Tropical cyclones are the deadliest disaster in the Bay of Bengal because it is accompanied by heavy precipitation and strong winds. Climate change can intensify tropical storms and cyclones in the North Indian Ocean coast, precisely the Bay of Bengal (BOB) region (Reddy et al. 2021). It is predicted to incur severe damage in Bangladesh. It also affects the major densely populated Indian coastal cities like Chennai, Visakhapatnam, Bhuvaneswar, and Kolkata. The storm surges with water travelling upstream from the sea at a high speed, further surging through nearby canals and waterways. It also dismantles embankments and causes damages in far inland instantly. History reveals that most deadly Tropical Cyclones and storm surges have consistently occurred in the Bay of Bengal. The severity of the worst surges in the Bay of Bengal is established by the fact that it only constitutes 5–6% of global tropical cyclones but records 80–90% of fatalities (Needham et al. 2015). The frequent occurrences of such Hydro-Metrological hazards with the sharpened climate change have revitalised the debates on human security. Global climate change is the linchpin that alters the frequency of natural hazards in Bangladesh.

5.9 Cross-Border Migration and Vulnerability

Climate changes and environmental vulnerabilities further exacerbate Bangladeshis' decision to move across the border. Panda argues that "vulnerabilities are interdependent through the mechanisms that increase exposure or sensitivity, as well as the processes that affect capacities" (Panda 2010). The topographical traits unique to Bangladesh increase the population's susceptibility to natural disasters and migration as a coping mechanism. There are other complex and interdependent effects brought on by changing climatic circumstances outside its boundaries. Internal displacement due to natural disasters is also spontaneously increasing in Bangladesh. However, Internally Displaced Persons (IDPs) have no legal or social recognition by the national governments or international forums. The influx of climate migrants to India from the environmentally vulnerable location of Bangladesh is alarming. People crossing the Indian border are often illegal and unauthorised (Mahmud 2023).

According to Chari's conceptualisation, the refugees are determined as.

> The state and/or powerful communities within a state are generally responsible for creating conditions which compel people to seek refuge across borders. These conditions include internal conflicts arising from a variety of reasons generically reflecting the failure of governance by the state. These conditions may be distinguished from population movements caused by natural calamities and environmental degradation, or those arising as the result of developmental policies pursued by the state, or those population movements motivated by the understandably human desire of people to pursue a better quality of life" (Chari 2003).

Due to climate change, India and Bangladesh would be indirectly impacted by changes to the bio-physical environment on either side of the border (Panda 2010) which could induce unintended migration from Bangladesh to India. Disasters like cyclones, floods and salinity increases in the southwest coast of Bangladesh, including the Sundarbans region. This has higher chances of people migrating to India. Residents in the border districts between India and Bangladesh experience migration as a means of adaptation and it is believed that India would be the most convenient country to move to. Flooding risk indicators are more closely linked with cross-border migration to India. Simultaneously, an increase in salinity triggers a significantly higher population exodus from Bangladesh's Bagerhat and Satkhira districts to India (Chen and Mueller 2019).

Asian News International (ANI) reported that as many as 9,233 Bangladeshis have been trying to enter India since 2019. From 1 January to 28 April 2022 alone, 4896 Bangladeshi nationals were held at the India-Bangladesh border while trying to enter India. Nearly 14,000 Bangladeshi nationals have been sent back and averted from entering the Bangladesh Border in India since 2019 (*ANI, April 29, 2022).* Most unauthorised immigrants who cross into India come from Bangladesh's southwest coast (Chen and Mueller 2019). This region of Bangladesh has a significant rise in soil contamination by saltwater and is unsuitable for agriculture. These people commonly cross the border illegally for work and shelter when internally displaced due to the loss of agricultural land or a tropical cyclone. They are frequently classified as economic refugees rather than climate refugees. It is alarming for India as no sound refugee policy framework exists to overcome the climate refugee crisis. The Government of India has initiated to curtail illegal migration through protected border laws and regulations like the Illegal Migrants Act 1983, the Foreigners Act 1946, and the Passport Act 1967 and also began constructing barbed wire fences along the heavily trafficked border. Furthermore, the Govt of India issued a National Register of Citizens (NRC) to regulate cross-border migration from Bangladesh (Dutta 2018). Nevertheless, the implementation of these regulations has been perplexing so far.

Sometimes, vulnerabilities are interdependent from the exposure and depend on the sensitivities of the ecosystem and its capacity to respond. The vulnerability of Bangladesh's ecosystem, people, and economy can correlate with the interdependence of shared resources of both India and Bangladesh. In the present situation, Bolstering Bangladesh's capacity owing to future climate catastrophes is thus important to prevent the problem of escalating climate refugees. Otherwise, it could double the burden on India due to such migration from Bangladesh.

5.10 Institutional Maladaptation and Hydro Diplomacy

In South Asia, discussions on water are frequently extremely political and influenced more by local and national interests than common regional concerns. Water management itself is overwhelmingly technical in its approach. Across the region, scientists and "hydrocrats" have dominated policy formulation and implementation for years. Information on water resources is closely guarded, with river flows sometimes considered state secrets. Social and ecological perspectives are underrepresented, and the public has limited opportunities to sufficiently understand issues or advocate on their own behalf.

The U.S. Senate report echoes that water scarcity leads to human security and conflicts across South Asia and Central Asia, where transboundary water resources are highly sensitive (U.S Senate 2011). The 'Hydropolitics' poses one of the major challenges among the neighbouring countries when unbalanced action of one country on water resources (may) affect its neighbours. Chinese water diversion through dam construction in the Yarlung Tsangpo (Upper stream of Brahmaputra) would impact the developmental processes of India's Northeastern States, such as Arunachal Pradesh. Further, the situation would become even more difficult as the water goes downwards to Bangladesh (Vishwanath 2018). The sufferers are the lower riparian states, which remain powerless to negotiate with the enormous upper riparian. The Political Economy of Indo-Bangladesh riparian communities have emerged with controversy over the potential environmental sustainability, the possibility of agricultural degradation, and water insecurity from the Tipaimukh Dam in India which is around 210 kms upstream from Bangladesh (Chellaney 2014). It has added to another interstate conflict. Protest against this dam, termed "environmentalism of the poor," transcended the frontier, and included such protesting groups in the Northeast region of India (Islam and Al-Amin 2019).

Climate change is an additional burden to the already fragile bilateral-trilateral cooperation of major riparian states in South Asia under the scenario of intensified water scarcity. The strengthening of the institutional framework can produce climate knowledge and preparedness across borders and enhance the binational regions' adaptive capacity and resilience (Wilder et al. 2010). Regarding bilateral climate change adaptation, the institutional capacity of all affected countries indeed matters. Shared interests for the common good have yet to be derived so far precisely because of the domination of national and local interests over regional interests.

5.11 Results and Discussion

Transboundary climate risks emanate on two grounds: (a) when climate change impact in one country generates risks to the people or economy in another country, and (b) an impact from one's adaptation in one country generates risks to people in another country. When countries fail to understand or manage the climate risks that arise outside their borders, which have spillover effects, it is considered a significant constraint towards adapting to cascading risk. It is essential to recognise that adaptation can also have cross-border effects, redistributing rather than reducing the vulnerability. For instance, the construction of embankments in the upstream catchments reduces the capacity of the flood plains to store water. Additionally, the area is made more delicate due to inappropriate maintenance of embankments.

Hence, the countries spur little bilateral regional or international cooperation on adaptation with the view of losing their sovereignty over shared resources and mandate. For instance, the meeting of the ministerial-level joint rivers commission of India and Bangladesh was held in 2022 after a long gap of 12 years (GoI 2022a, b). However, integrating national adaptation plans with regional adaptation measures will bring mutual benefits by supporting the capacity of the member countries through cooperation. Because "particularly for the continental countries of the region, unilateral actions cannot be fully effective on their own because the environmental conditions that produce the threats do not stop at borders" (ADB 2014).

Regional and global integration in terms of trade and connectivity has immensely benefited the countries in the Hindu Kush Himalayas (HKH) region to revoke their poverty, increase the standard of living and move across the region. However, regional integration has also brought distinct risks to these countries due to underdeveloped risk management systems and the globalisation of local risk with greater exposure to global risks. With greater dependence on transboundary natural resources, the natural resources are receiving immense pressure due to poorly developed natural resource governance and regional integration.

The challenges that emerge upon mitigating transboundary climate risks include:

1. Limited acknowledgement of adaptation as a regional or global good.
2. Limited institutional mechanisms and country-level collaboration for exchanging risk information.
3. Highly fragmented risk assessments that fail to recognise connections with other sectors, as a result of which hazards are only partially understood, and risk communication is ineffective.
4. Insufficient thorough and integrated climate risk evaluations.
5. Geopolitical circumstances and the reluctance to provide information about potential risks that impede bilateral or regional collaboration.
6. Criticality of having cross-border cooperation among scientific, civil society and government stakeholders.

5.12 Conclusion

Climate denotes the presence of a formidable force. It is not a local or national effect demarcated by the people but a global effect. For instance, cyclone is a specific activity which has a global impact. It originates in a micro space and spreads across a wide area spatially and temporally due to the coercion of thermal and physical reactions. Humans have no strength to stop or control the cyclone or such sudden-onset events. But humans can control or reduce its devastating impacts and implications by promptly coordinating prediction, warnings, and adaptive mechanisms.

The primary constraints in the regional groupings for climate cooperation are concerned with the location for holding climate talks within the region. Secondly, sometimes countries are reluctant to share adequate data, and shared ones are unreliable due to a lack of authenticity and accuracy. Scientific predictions must be the guiding tool, and socio-economic coordination and capacity building must be included in the adaptation framework. However, it is impossible to be secured from climate risks, if states adapt to climate change impacts within their territorial borders, as several of its impacts are transboundary. Bi-lateral and multi-lateral policy frameworks could open up to deepen and integrate their agenda, knowledge, and readiness towards ensuring a just and equitable share of water resources and its benefits. It must also bring the people-centred paradigm of inclusiveness, decision making and capacity-building across the borders. In South Asia, climate change can become a constant thread for cooperation and conversation between the countries on shared natural resources despite the socio-political differences.

References

ADB (2014) Assessing the costs of Climate change and Adaptation in South Asia. Asian Development Bank, Philippines

Adger et al (2009) Nested and teleconnected vulnerabilities to environmental change. Front Ecol Environ, 7(3):150–157. https://pubmed.ncbi.nlm.nih.gov/32313513/

Ahmed et al (1999) Vulnerability of forest ecosystems of Bangladesh to climate change. In: Huq et al (eds) Vulnerability and Adaptation to Climate change for Bangladesh. Kulwer Academic Publishers

Ali A (1999) Climate change impacts and adaptation assessment in Bangladesh. Climate Res 12:109–116. https://doi.org/10.3354/cr012109

ANI (2022) More than 14,000 Bangladeshi nationals sent back since 2019. ANI. Retrieved October 6, 2023, from https://www.aninews.in/news/national/general-news/more-than-14000-bangladeshi-nationals-sent-back-since-201920220429140056/

Anisimova and Magnan (2023) The global transboundary climate risk report. The Institute for sustainable development and international relations & adaptation without borders. https://adaptationwithoutborders.org/

Banda ML (2018) Global adaptation law: Optimizing legal design for multi-level public goods after the Paris Agreement. Vanderbilt J Transnatl Law 51(4):997–1026

Benzie M, Persson S (2019) Governing borderless climate risks: moving beyond the territorial framing of adaptation. Int Environ Agreem: Polit, Law Econ 19(4–5):369–393. https://doi.org/10.1007/s10784-019-09441-y

Benzie M (2014) National Adaptation Plans and the indirect impacts of climate change, Policy Brief. Stockholm Environment Institute, Stockholm, Sweden. https://mediamanager.sei.org/documents/Publications/Climate/SEI-PB-2014-Indirect-climate-impacts-NAPs.pdf

Benzie M et al (2016) Introducing the transnational climate impacts index: indicators of country-level exposure-methodology report. Working Paper 2016–07. Stockholm Environment Institute, Stockholm, Sweden. https://www.sei.org/publications/transnational-climate-impacts-index/

Pramanik, Bhaduri (2016) Impact of climate change on water resources in Indian Himalaya. In Bisht JK et al (eds) Conservation agriculture. Springer, Singapore

Biswas AK (2011) Cooperation or conflict in transboundary water management: case study of South Asia. Hydrol Sci J, 56 (4):662–670. https://doi.org/10.1080/02626667.2011.572886

Carter TR, Benzie M, Campiglio E, Carlsen H, Fronzek S, Hildén M, Reyer CP, West C et al (2021) A conceptual framework for cross-border impacts of climate change. Glob Environ Chang 69:102307. https://doi.org/10.1016/j.gloenvcha.2021.102307

Challinor et al (2018) Transmission of climate risks across sectors and borders. Phil Trans R Soc A 376(2017030):1–23. https://doi.org/10.1098/rsta.2017.0301

Chari PR (2003) Refugees, migrants and internally displaced persons in South Asia: An overview. In: Chari PR, Joseph M, Suba Chandran D (eds) Missing boundaries: Refugees, migrants, stateless and internally displaced persons in South Asia. Manohar Publishers, New Delhi, pp 17–38

Chellaney B (2014) Water, power, and competition in Asia. Asian Surv 54(4):621–650

Chen and Mueller (2019) Climate-induced cross-border migration and change in demographic structure. Popul Environ 41:98–125. https://doi.org/10.1007/s11111-019-00328-3

Das and Bandyopadhyay (2015) Climate change adaptation in the framework of regional cooperation in South Asia. Carbon & Clim Law Rev 9(1):40–54

Dutta A (2018) Political destiny of immigrants in Assam: national register of citizens. Econ Pol Wkly 53(8):18–21

Eakin H et al (2008) Nested vulnerability: exploring cross-scale linkages and vulnerability teleconnections in Mexican and Vietnamese coffee systems. Environ Sci Policy. https://doi.org/10.1016/j.envsci.2008.09.003

Ercins et al (2021) Cross-border climate vulnerabilities of the European Union to drought. Nat Commun 12(3322):1–9. https://doi.org/10.1038/s41467-021-23584-0

Galaz V, Galafassi D, Tallberg J, Boin A, Hey E, Ituarte-Lima C, Westley F et al (2014) Connected risks, connected solutions. Stockholm Resilience Centre, Stockholm University, and the Global Challenges Foundation, Stockholm

Goldin I, Mariathasan M (2014) The butterfly defect: how globalization creates systemic risks, and what to do about it. Princeton University Press, Princeton

Government of India (2011) Fact sheet on agreements and other MOUs signed between India and Bangladesh, in Ministry of External Affairs. https://www.mea.gov.in/bilateraldocuments.htm?dtl/5229/Fact+Sheet+on+Agreemens+and+other+MOUs+signed+between+India+and+Bangladesh. Accessed 23 Mar 2023

Government of India (2022) Farakka Barrage Project, Farakka, in Ministry of Water. https://jalshakti-dowr.gov.in/farakka-barrage-project-farakka/. Accessed 14 July 2023

Government of India (2022) 38th Meeting of Ministerial level Joint Rivers Commission of India and Bangladesh held at New Delhi, in Press Information Bureau. https://www.pib.gov.in/PressReleseDetailm.aspx?PRID=1854557. Accessed 19 July 2023

Hanasz P (2014) Power flows: hydro-hegemony and water conflicts in South Asia. Security Challenges 10(3):95–112

Haque et al (2018) Present and future fluvial, tidal and storm surge flooding in coastal Bangladesh. In: Nicholls RJ et al. (eds) Ecosystem services for well being in Deltas. Palgrave Macmillan. https://doi.org/10.1007/978-3-319-71093-8_16

Huda MS (2020) Energy cooperation in South Asia: utilizing natural resources for peace and sustainable development, 1st edn. Oxon, Routledge

ICIMOD (2009) The changing himalayas: impact of climate change on water resources and livelihoods in the Greater Himalayas. International Centre for Integrated Mountain Development, Nepal

IPCC (2001) Climate Change 2001: The scientific basis. contribution of working group I to the third assessment report of the intergovernmental panel on climate change. In: Houghton JT et al (eds). Cambridge University Press, Cambridge, United Kingdom and New York, NY, USA

IPCC (2014) Climate Change 2014: Impacts, adaptation, and vulnerability. Part A: Global and sectoral aspects. Contribution of Working Group II to the Fifth Assessment Report of the Intergovernmental Panel on Climate Change [Field CB, Barros VR, Dokken DJ, Mach KJ, Mastrandrea MD, Bilir TE, Chatterjee M, Ebi KL, Estrada YO, Genova RC, Girma B, Kissel ES, Levy AN, MacCracken S, Mastrandrea PR, White LL (eds)]. Cambridge University Press, Cambridge
Islam MN, Al-Amin M (2019) The rampal power plant, ecological disasters and environmental resistance in Bangladesh. Int J Environ Stud 76(6):922–939. https://doi.org/10.1080/00207233.2019.1662183
Leichenko et al (2010) Climate change and the global financial crisis: a case of double exposure. Ann Assoc Am Geogr 100(4):963–972
Lesnikowski A, Ford J, Biesbroek R, Berrang-Ford L, Heymann SJ et al (2016) National-level progress on adaptation. Nat Clim Chang 6(3):261–264. https://doi.org/10.1038/nclimate2863
Mahmud H (2023) International migration in Bangladesh: A political economic overview. In: Irudaya Rajan S (eds) Migration in South Asia, IMISCOE Regional Reader. Springer
Mirumachi N (2013) Trans-boundary water security: Reviewing the importance of national regulatory and accountability capacities in International transboundary river basins. In: Lankford B (eds) Water security: principles, perspectives, and practices. Routledge, Oxon
Miyan MA (2015) Droughts in Asian least developed countries: vulnerability and sustainability. Weather Clim Extrem 7:8–23. https://doi.org/10.1016/j.wace.2014.06.003
Molden D, Sharma E, Shrestha AB, Chettri N, Pradhan NS, Kotru R et al (2017) Advancing regional and transboundary cooperation in the conflict-prone Hindu Kush-Himalaya. Mt Res Dev 37(4):502–508. https://doi.org/10.1659/mrd-journal-d-17-00108.1
Needham HF, Keim BD, Sathiaraj D et al (2015) A review of tropical cyclone-generated storm surges: Global data sources, observations, and impacts. Rev Geophys 53(2):545–591. https://doi.org/10.1002/2014rg000477
Nepal S, Shrestha AB (2015) Impact of climate change on the hydrological regime of the Indus, Ganges and Brahmaputra River basins: a review of the literature. Int J Water Resour Dev 31(2):201–218. https://doi.org/10.1080/07900627.2015.1030494
Pal S, Talukdar S, Ghosh R (2020) Damming effect on habitat quality of riparian corridor. Ecol Indic, 114(106300).
Panda A (2010) Climate-induced migration from bangladesh to india: issues and challenges. SSRN Electron J. https://ssrn.com/abstract=2186397 or https://doi.org/10.2139/ssrn.2186397
Pandey P (2014) Bangladesh, India and Fifteen years of peace Future directions of the Ganges Treaty. Asian Surv 54(4):651–673. https://doi.org/10.1525/as.2014.54.4.651
Pescaroli, Alexander (2015) A definition of cascading disasters and cascading effects: Going beyond the "toppling dominos" metaphor. Planet@Risk, 3(1):58–67
Prabhakar SVRK, Shivakoti BR, Corral AF et al (2018) Transboundary impacts of climate change in asia: making a case for regional adaptation planning and cooperation. IGES Discussion Paper. Institute for Global Environmental Strategies and Global Development Network. Hayama, Japan. https://www.iges.or.jp/en/pub/transboundary-impacts-climate-change-asia/en
Reddy PJ et al (2021) Impact of climate change on intense Bay of Bengal tropical cyclones of the post-monsoon season: a pseudo global warming approach. Clim Dyn 56:2855–2879. https://doi.org/10.1007/s00382-020-05618-3
Rudra K (2018) Rivers of the ganga Brahmaputra—Meghna Delta: A fluvial account of Bengal. Springer, Switzerland
Sabin TP et al (2020) Climate change over the Himalayas. In Krishnan R et al (eds) Assessment of climate change over the Indian region. Springer eBooks. https://doi.org/10.1007/978-981-15-4327-2_11
Saklani et al (2020) Hydroenergy cooperation in South Asia: Prospects for transboundary energy and water security. Environ Sci Policy 114:22–34
Salimzadeh, Kalantari (2015) Agricultural water resources limitations and their effects on the socioeconomic conditions of wheat farmers: A case study of Maku City in Iran. In Hipel et al. (eds) Conflict resolution in water resources and environmental management. Springer, Switzerland

SAWI (2019) Annual Report July 2019—June 2020, South Asia Water Initiative, World Bank. https://documents1.worldbank.org/curated/en/720411611563614917/pdf/South-Asia-Water-Initiative-SAWI-Annual-Report-for-July-2019-June-2020.pdf
Scott CA, Varady RG, Meza F, Montaña E, de Raga GB, Luckman B, Martius C et al (2012) Science-policy dialogues for water security: addressing vulnerability and adaptation to global change in the arid Americas. Environ: Sci Policy Sustain Dev, 54(3):30–42. https://doi.org/10.1080/00139157.2012.673454
Sullivan (2011) Glaciers, monsoons, rivers, and conflict: China and South Asia. Georg J Int Aff, Winter/Spring 12(1):107–117
US Senate (2011) Avoiding water wars: water scarcity and central asia's growing importance for stability in Afghanistan and Pakistan. A Majority Staff Report, Committees on Foreign Relations United States Senate, Washington DC. https://www.foreign.senate.gov/publications/download/s-prt-112-10-avoiding-water-wars-water-scarcity-and-centrals-asias-growing-importance-for-stability-in-afghanistan-and-pakistan
Vishwanath A (2018) Paddling upstream: transboundary water politics in south Asia. Carnegie Endowment for International Peace, pp 1–6. CarnegieIndia.org
Wilder M, Scott CA, Pablos NP, Varady RG, Garfin GM, McEvoy J et al (2010) Adapting across boundaries: climate change, social learning, and resilience in the U.S.–Mexico border region. Ann Assoc Am Geogr 100(4):917–928. https://doi.org/10.1080/00045608.2010.500235

Chapter 6
Assessing the Efficacy of Glacier Inventories to Evaluate Climate Change Impacts: Key Takeaways from Baspa River Basin

Lydia Sam, Anshuman Bhardwaj, Shaktiman Singh, Benjamin C. Sam, and Rajesh Kumar

Abstract A glacier inventory is helpful in studying temporal glacier changes, glacio-hydrological regimes, future sea level rise, and climate model optimisation for different scenarios. The compilation of any glacier inventory requires substantial manual and computational efforts. However, the inaccuracies in glacier inventories have implications for modelled glacio-hydro-climatological results, necessitating a need to understand the degree of area uncertainties in the input inventories. In this work, we first developed a glacier inventory dataset using high-resolution images and field validations for our case study site, i.e., Baspa River Basin, India. Subsequently, through spatial comparison, we estimated the extent of area uncertainties across the available regional and global-scale glacier inventories. These area uncertainties are found to be significantly high, within a range of $\pm$ 23%, and they can further magnify, if the mapped basin area has a higher proportion of debris-covered glaciers. We further performed a sensitivity analysis to assess the impact of area discrepancies on a glacio-hydrological model outcome. The change in debris-covered glacier area by $\pm$ 25% resulted in alteration of average monthly discharge by up to $\pm$ 16%. It is significant enough to highlight the need for quality-controlled inventory data for running such models. In the last part of this chapter, we present our recommendations

L. Sam (✉) · A. Bhardwaj · S. Singh
School of Geosciences, University of Aberdeen, Aberdeen, UK
e-mail: lydia.sam@abdn.ac.uk

A. Bhardwaj
e-mail: anshuman.bhardwaj@abdn.ac.uk

S. Singh
e-mail: shaktiman.singh@abdn.ac.uk

B. C. Sam
Department of Natural and Applied Sciences, TERI School of Advanced Studies, Delhi, India

R. Kumar
Department of Environmental Science, Central University of Rajasthan, Ajmer, Rajasthan, India
e-mail: rajesh.kumar@curaj.ac.in

A. Sarkar et al. (eds.), *Risk, Uncertainty and Maladaptation to Climate Change*,
Disaster Risk Reduction, https://doi.org/10.1007/978-981-99-9474-8_6

for implementing the probable uncertainty scenarios while discussing the research on the future projections. While we have reached a critical point of climate change in human history, there are existing uncertainties in assessing future water security due to various data and geopolitical reasons. The perspectives offered in this chapter can help better account for glacier area-related uncertainties in glacio-hydrology.

Keywords Glacier inventory · Mapping uncertainty · Climate change · Hindu Kush-Himalaya · Debris-covered glacier

6.1 Introduction

Glaciers are well-established climate change indicators, and their continuous monitoring is imperative for understanding the complexities of glacio-climatic interactions (Barry 2006; Bhardwaj et al. 2016a, b; Sam et al. 2019; Shekhar et al. 2017; Singh et al. 2016a). The awareness on glacier monitoring has persistently increased since 1990 after the Intergovernmental Panel on Climate Change (IPCC) started to include glacier fluctuation data in their assessments as an indicator (Barry 2006). However, the importance of glaciers as climate indicators was first recognized in the latter half of the nineteenth century. Figure 6.1 depicts a timeline highlighting several events that established the glaciological sciences as we know them today and led to the formation of the World Glacier Monitoring Service (WGMS) (Radok 1997). The utility of glacier inventories to temporarily monitor global glaciers soon became apparent (WGMS 1989). A glacier inventory became necessary to study the past, present, and future of the global and regional ice-water budget and sea level changes. A digital version of the World Glacier Inventory (WGI), which began with an initial global coverage of 25% in 1995 (Bedford and Haggerty 1996) increased to 48% in 2009 (Cogley 2009) that is available from the National Snow and Ice Data Centre (NSIDC) (Pfeffer et al. 2014). However, the incompleteness of the WGI and the lack of glacier outlines limited its applicability and prompted subsequent global and regional initiatives such as the Global Land Ice Measurements from Space (GLIMS) inventory in 1995 (Raup et al. 2007), the Randolph Glacier Inventory (RGI) in 2011 (Pfeffer et al. 2014), the Chinese Glacier Inventory (CGI) in 2002 (Guo et al. 2015; Shi et al. 2010), the Geological Survey of India (GSI) inventory in 1999 (Kaul 1999; Raina and Srivastava 2008), the International Centre for Integrated Mountain Development (ICIMOD) inventory in 2001 (Bajracharya and Shrestha 2011; Mool et al. 2001), and the Glacier Area Mapping for Discharge from the Asian Mountains (GAMDAM) Inventory (GGI) in 2014 (Nuimura et al. 2015). These inventories act as direct input to various geophysical models, such as the ones that project glacier changes (Hagg et al. 2013; Hirabayashi et al. 2013; Kumar et al. 2015; Möller and Schneider 2010), glacio-hydrological regimes (Bliss et al. 2014; Immerzeel et al. 2010; Kaser et al. 2010; Lutz et al. 2016; Prasch et al. 2013; Radić and Hock 2013),

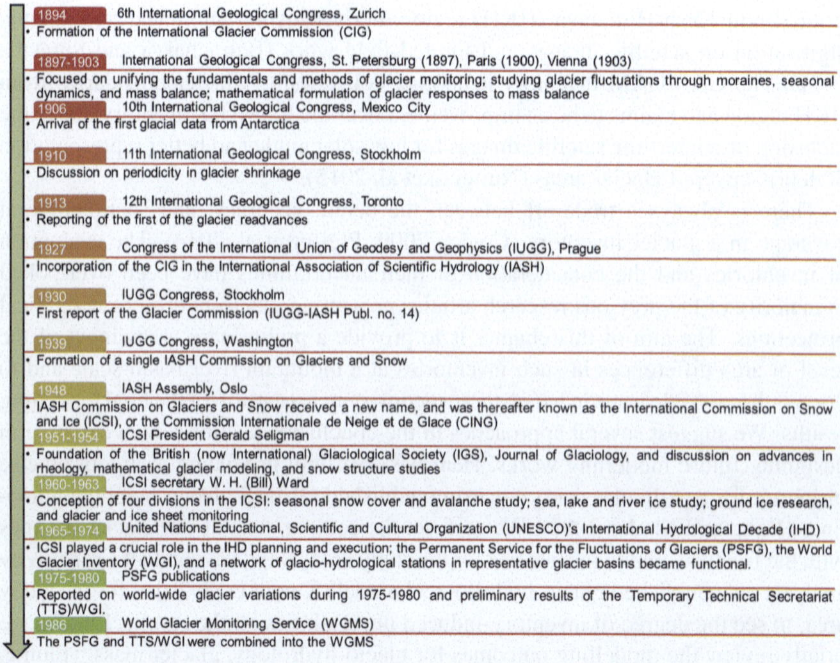

Fig. 6.1 Timeline of events leading to the formation of the World Glacier Monitoring Service (WGMS)

climate regimes (Sakai et al. 2015), and sea level rise (Berthier et al. 2010; Arendt et al. 2002; Gardner et al. 2013; Marzeion et al. 2012; Moore et al. 2013; Radić and Hock 2011).

The GLIMS inventory (GLIMS Consortium 2005), although incomplete in global coverage (~58%), has extensive set of attributes and includes multitemporal boundaries of several thousands of glaciers (Raup et al. 2007). The RGI is a globally complete inventory with uncertainty assessment (±5%) estimated through analysis of single glacier and basin-scale uncertainties (Pfeffer et al. 2014). The second version of CGI is the first inventory using references from Differential Global Positioning System (DGPS) observations for 23 glaciers and digitisation on high-resolution satellite imageries for improving the overall accuracy (within ± 3.2%) (Guo et al. 2015). However, the accuracy of CGI further diminishes for debris-covered glacier area (±17.6%) and covers ~ 86% of the total glacierised area of China (Guo et al. 2015). The GSI inventory is largely obtained from past topographic maps and aerial and field photographs giving tabular information on different topographical characteristics of individual glaciers, but due to its unavailability in digital format, its accuracy and uncertainty has never been estimated (Kaul 1999; Raina and Srivastava 2008). The ICIMOD inventory is a complete collection of glacier boundaries and ice thickness from Hindu Kush-Himalaya (HKH) region, derived using automated and manual digitisation on satellite images and limited field work (Bajracharya and Shrestha 2011). The GAMDAM inventory is another collection of glacier boundaries from HKH region and claims to be an improvement over the ICIMOD inventory due to the inclusion of wintertime satellite images for low solar angle and better representation of debris-covered glacier areas (Nuimura et al. 2015).

There is always a trade-off between the detail, accuracy, and spatiotemporal coverage in a glacier inventory (Cogley 2009; Pfeffer et al. 2014). The evaluation of inventories and the consideration of their uncertainties have been overlooked in majority of the previous research articles presenting glacio-hydro-climatological projections. The aim of this chapter is to provide a preliminary assessment of the level of area differences in such inventories at a mountain river basin-scale and to discuss the possible ways in which these inconsistencies can affect the final modelling results. We suggest several approaches in the conclusion section which can help in designing future modelling works. Here we start with producing a largely precise inventory for a particular study site using published GLIMS glacier mapping guidelines (Raup and Khalsa 2007). We perform the comparisons of our glacier outlines with the freely available inventories for the study area. Next, we conduct a sensitivity analysis on a published glacio-hydrological model (Kumar et al. 2016) for the study area, to see the degree of inventory-induced uncertainties on the results. Finally, we briefly review the modelling outcomes for glacio-hydrology, glacier mass, volume, climate change effects, and sea level in view of the probable area misestimates. As we have reached a critical point of anthropogenic climate change, the perspectives offered in this chapter can help better account for glacier area-related uncertainties while assessing future water availability in glacierised mountains.

6.2 Materials and Methods

The glaciers of Baspa River Basin have been extensively studied for glacier dynamics (Sam et al. 2016, 2018), glacier lakes (Bhardwaj et al. 2015b), glacier landform mapping (Bhardwaj et al. 2015a, 2016d), runoff estimations (Kumar et al. 2016; Singh et al. 2018), and glacier sediment load (Kumar et al. 2018). We generated our own glacier vector outlines for the ~1100 km^2 Baspa River Basin in the Western Indian Himalaya (Fig. 6.2).

6.2.1 *Rationale for Selecting the Study Site*

Our previous research and familiarity with the Baspa River Basin prompted us to select it for the high-accuracy mapping of glaciers and to subsequently use those digital outlines for a comparison with several existing inventories. The Baspa River is ~ 70 km long and is a significant contributor to the Sutlej River in its upper

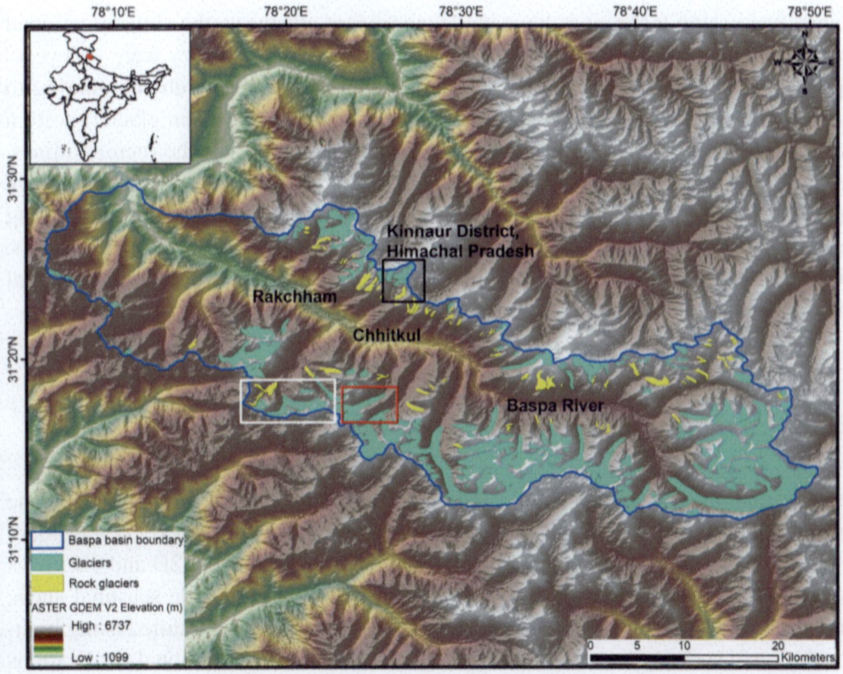

Fig. 6.2 Glaciers in the Baspa River Basin (topography: hill-shaded Advanced Spaceborne Thermal Emission and Reflection Radiometer Global Digital Elevation Model Version 2 (ASTER GDEM V2), METI-NASA). The white, red, and black rectangles provide contextual information for Figs. 6.5 and 6.6a, b, respectively. The inset map indicates the location of the Baspa Basin (red outline) in India (black outline)

course. This river basin is located in the Kinnaur District in the Indian state of Himachal Pradesh and is heavily glacierised (Fig. 6.2) with elevations in the range of ~ 1,100–6,700 m above the sea level (asl) while the glaciers are located within the altitudinal range of ~ 4,100 m–6,450 m asl with the mean elevation of ~5,200 m. Based on an inventory in 1970s (Vohra 1980), the GSI reported 89 glaciers within the river basin (Raina and Srivastava 2008), and a report prepared by Vohra (1980) for the International Association of Hydrological Sciences (IAHS) described the problems and issues encountered by the GSI during an aerial photograph-based glacier inventory in the Himalaya by citing the example of the Baspa Basin glacier inventory. Vohra (1980) highlighted several prominent problem areas encountered while mapping the Baspa Basin glaciers, such as, delineation of the snow line (due to seasonal snow), estimation of glacier length (due to ice-cored, and arcuate or curved terminal moraines), quantification of glacier depth (in accumulation zones), identification of glacier termini (obscured by thick debris cover, or multiple termini due to their separation by medial moraines), exclusion of snow patches on the valley walls from the glacier boundaries (due to mountain shadows), and mapping of supraglacial debris (due to its resemblance with periglacial debris). Although remote sensing and image processing techniques have made significant advances, the abovementioned mapping issues are nearly as difficult to address even today as they were ~40 years ago (Bhardwaj et al. 2014, 2016a; Racoviteanu et al. 2009), thereby amounting to uncertainties in the glacier inventories. Therefore, the Baspa Basin glaciers were a suitable choice for performing an assessment of the uncertainty in the vector outlines, as these glaciers possess all of the possible geomorphological elements (i.e., thick supraglacial debris cover, branched accumulation zones separated by debris, hanging snow patches, periglacial landforms such as ice-cored moraines and rock glaciers, and steep-walled cirques hidden by mountain shadows in the satellite images) that can hinder the interpretation and mapping using remote sensing images.

6.2.2 Mapping of the Glaciers in Baspa River Basin

For this mapping, we took a simple and effective approach as shown in Fig. 6.3. The entire glacier outline mapping process was divided into two phases: (1) a primary phase in which the glacier outlines were manually digitized using 2D and 3D views generated from Landsat 8 imagery (20th August 2014) with little seasonal snow, and (2) a corrective phase in which the digitized outlines were modified using high-resolution 3D perspective views from Google Earth (GE) and additional verifications from field observations. We opted for manual digitisation because our glaciers were majorly debris-covered, and the outlines had to be considerably accurate in order to perform uncertainty assessment of other inventories. We have essentially followed the well-accepted and published GLIMS glacier mapping guidelines (Raup and Khalsa 2007) for mapping the glaciers. To ensure considerable accuracy in mapping, we combined inputs from high-resolution satellite images and digital elevation models (DEMs) with subsequent corrections of the glacier boundaries using field verifications. Thus, the present mapping is performed at multiple scales, starting with Landsat resolutions, followed by Google Earth (GE) high-resolution observations, and subsequent field-based modifications. While performing 2D mapping, using simultaneous inputs from 3D perspective views by draping satellite imagery over DEMs can improve the manual digitization process immensely (Raup et al. 2007). This approach was helpful in glacier snout/tongue identification, bergschrund (i.e., crevasse along the glacier headwall) identification, and differentiating supraglacial debris from periglacial debris on moraine slopes. We use our glacier outlines as the base dataset to perfom an uncertainty analysis of the existing inventories which are prepared by various analysts using different mapping approaches. A point worth mentioning here is that another independent mapping study (Mir et al. 2017) for the Baspa River Basin has reported very similar number and area estimates for the glaciers in this basin as compared to our base inventory (henceforth referred to as BI).

The primary phase started by downloading the Landsat 8 scene of 20 August 2014 (scene ID: LC81460382014232LGN00) that covered the entire Baspa Basin and was mostly free of seasonal snow and clouds over the glacierised regions. We generated

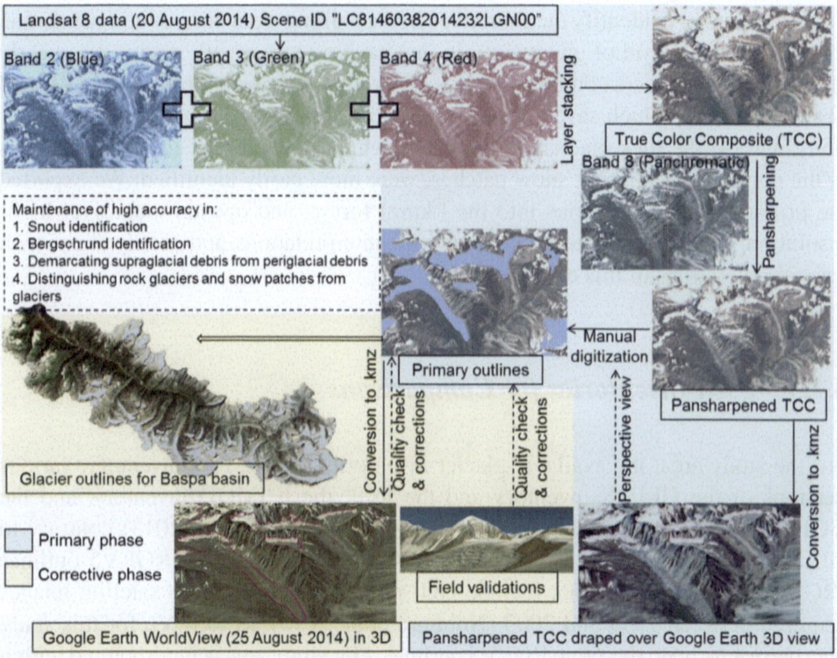

Fig. 6.3 Glacier outline mapping procedure using Landsat 8 data, GE images, and perspective views

a true colour composite (TCC) image at a spatial resolution of 30 m using bands 2 (blue), 3 (green), and 4 (red), as a TCC view is comparable with high-resolution GE images. However, we additionally generated a false colour composite (FCC) using bands 3 (green), 4 (red), and 5 (near infrared or NIR) for improved visualisation of glacier features. We performed pansharpening on the generated TCC and FCC images using the panchromatic band 8 in ArcGIS version 10.4, which improved the spatial resolution of the TCC and FCC images to 15 m. The ArcGIS Pansharpening tool incorporates the Gram-Schmidt method (Maurer 2013) with predefined tested weights for various bands of Landsat 8 data. The pansharpened TCC image was exported to a.kmz file extension and opened in GE to facilitate 3D visualisation. We adopted a "terminus-to-terminus" manual mapping approach (Nagai et al. 2016) to first delineate the outer polygon representing the boundary for each glacier. Afterwards, we removed the rocky outcrops and nunataks, which were not a part of the typical supraglacial debris or moving glacier bodies, from the digitised outlines.

The available GE images were acquired on 25th August 2014, which was very close to the date of acquisition of the Landsat 8 image used in this study. This greatly improved our mapping in several ways: (1) the nearer dates of the image acquisitions provided us with an opportunity to modify the medium-resolution mapping results using high-resolution images with similar glacier facies and periglacial environments; (2) similar glacier and periglacial environments in the high-resolution images helped us to identify the sources of error resulting from the medium-resolution mapping; (3) the level of uncertainty in the mapping was greatly reduced as nearly all of the basin glaciers could be clearly observed in the submeter resolution 2D and 3D views in GE, which saved considerable time and effort during our field validations; and (4) the rock glaciers in the basin, which often appear similar to glaciers in the presence of adjacent snow patches, were more easily identified. We exported the primary glacier polygons into the (.kmz) format and opened them over high-resolution images in GE to identify any misinterpretations and rectify them with several reiterations of this entire exercise.

6.2.3 Used Inventories for Comparisons

For the study area, the available glacier inventories are the GSI inventory, various versions of the GLIMS inventory and the RGI, the ICIMOD inventory, and the GGI. The GGI excludes thin ice on headwalls (Nuimura et al. 2015) contrary to GLIMS guidelines (Raup and Khalsa 2007). Here, we opted for RGI V5 outlines (RGI Consortium 2015) that include GGI polygons derived from satellite images acquired between 1999 and 2003 (Nuimura et al. 2015), as RGI V6 for this study area carries forward the same RGI V5 outlines. Our study area is incorporated within the RGI Region 14 South Asia West of Version 5.0 (V5). We re-projected the RGI V5 glacier outlines from the Geographic coordinate system (GCS) World Geodetic System (WGS) 1984 to the WGS 1984 Universal Transverse Mercator (UTM) zone 44N coordinates. We used the 2012 GLIMS outlines (GLIMS and National Snow and Ice Data Centre 2005, updated 2012). We also used the ICIMOD inventory (Bajracharya and Shrestha 2011) derived from satellite images acquired between 2002 and 2008 (http://rds.icimod.org/Home/DataDetail?metadataId=9361). We also prepared a digital version of the GSI inventory, which is originally in a printed format (Raina and Srivastava 2008), for comparison purposes. This inventory is primarily based on visual interpretations of aerial photographs of the Indian Himalaya. We used the basin boundary given in the scanned GSI glacier inventory map and our basin boundary derived from ASTER GDEM V2 to perform georeferencing. The basin boundaries matched considerably at most of the corners, and we used those points as control points (42 in total) for georeferencing in ArcGIS software using first-order polynomial transformations, acceptable root mean square error (RMSE) limits (total RMSE = ~ 0.37), and cubic convolution resampling. The projection assigned during the georeferencing was the same as rest of the datasets, i.e., WGS 1984 UTM 44N. Finally, the polygons for the GSI inventory were digitised in ArcGIS software at sufficiently high magnification levels.

Table 6.1 Cumulative glacier counts and area statistics

Inventory	Number of glaciers	Total glacier area (km^2)	Minimum area (km^2)	Maximum area (km^2)	Mean area (km^2)	Common area in the BI (km^2)	OI with respect to the BI	Uncertainty (%) in area with respect to the BI
BI	112	186.58	0.06	32.68	1.67	186.58	1	0
GSI	89	227.87	0.07	35.92	2.59	164.76	0.80	+ 22.12
ICIMOD	148	158.52	0.02	29.94	1.07	147.37	0.85	−15.04
GLIMS2012	138	148.88	0.01	28.88	1.08	141.61	0.84	−20.20
RGI V5	139	160.26	0.03	30.79	1.15	150.28	0.87	−14.11

6.2.4 Area Uncertainty Estimation

We calculated the cumulative overlapping index (OI) for each of the inventories with respect to the BI (Table 6.1). This index, which was suggested by Nagai et al. (2016), is calculated using Eq. 6.1:

$$OI = \sqrt{\frac{OA * OA}{TA_{BI} * TA_X}} \quad (6.1)$$

where, OA is the overlapping glacierised area, and TA_{BI} and TA_X are the total glacierised areas in the BI and the test inventory, respectively. A higher OI (i.e., closer to 1) corresponds to a larger proportion of overlapping area between the two inventories and smaller over- or underestimations of the glacierised area in the test inventory.

6.2.5 Sensitivity Analysis for Glacio-Hydrological Model

A glacio-hydrological model is an important approach to estimate past and future discharge and mass balance of a glacierised catchment in the data deficient Himalayan region. There are mainly two types of glacio-hydrological models, namely, energy balance models and degree-day models (Hock 2003). Due to lesser requirements of observed data, computational ease, and considerable accuracy, temperature index-based degree-day glacio-hydrological models have been more extensively used in Himalaya for estimation of discharge (Arora et al. 2010; Kumar et al. 2016; Pradhananga et al. 2014; Li et al. 2016; Singh and Bengtsson 2003; Singh and Jain 2003; Singh et al. 2008, 2018) and catchment wide mass balance (Azam et al. 2014a; Kumar et al. 2016; Shea et al. 2015) in comparison to energy balance models (Azam et al. 2014b; Fujita and Ageta 2000; Fujita and Sakai 2014; Kayastha et al. 1999; Lejeune et al. 2013; Miles et al. 2016; Mölg et al. 2012; Shrestha et al. 2015). To assess the magnitude of changes in the output of a glacio-hydrological model caused by uncertainties in total glacierised area in general and areal estimates for debris-covered parts in particular, here we use a semi-distributed temperature index-based glacio-hydrological model validated and published for our study area (Kumar et al. 2016). Equation 6.2 numerically describes this model (Kumar et al. 2016).

The choice to use this model is based on the facts that it has already been validated and established for our study area and it is a generic model representing all degree-day models predominantly used to study Himalayan glacio-hydrology. We also had the data to validate the outcomes of the sensitivity analysis using this model.

$$M = DDF \times T_m \tag{6.2}$$

where,

M = melt (in mm for a given period of time).

DDF = Degree-day factor (in mm °C^{-1} day^{-1}).

T_m = Mean temperature (in °C).

The model uses different DDF values for snow, debris-free and debris-covered glacierised areas based on measurements carried out in the field. For debris-covered glacierised area, different DDF values were calculated and used depending on the altitudinal band due to the heterogeneity in debris-cover thickness. The model calculated melt (in mm) from snow, debris-free and debris-covered glacierised area caused by temperature which is added by liquid precipitation (in mm) and the melt caused by liquid precipitation on snow and debris-free ice surface (in mm). This runoff (in mm) from each land cover type is multiplied by their respective surface area and a conversion factor to obtain run-off in m^3 per unit time (for further details please refer to Kumar et al. 2016). Since the point run-off (in mm) obtained from such glacio-hydrological models for different land covers is integrated over the total surface area, any minute uncertainty in the surface area will introduce proportionally large uncertainties in the model outputs. The idea behind this sensitivity analysis was to find the range of uncertainty in estimated discharge corresponding to the range of uncertainty in the estimated glacier area.

6.3 Results and Discussion

6.3.1 Cumulative Glacier Counts and Area Uncertainties

The typical WGI and GLIMS inventory area cutoffs for an ice body that is to be categorised as a glacier is 0.01 km^2 (Pfeffer et al. 2014; Kargel et al. 2014). For our study area, all of the glaciers in the inventories used in this study showed an area $\geq$ 0.01 km^2. The cumulative glacier counts and area statistics for the employed inventories are given in Table 6.1. Figure 6.4 details the distribution of the glacier counts and cumulative areas for different sizes of area classes based on the total area according to each inventory. We performed this analysis by selecting most of the area classes similar to Nagai et al. (2016) for the Bhutanese Himalaya, ranging from 0.01 to $\geq 10\,km^2$. Although we observe area uncertainties in the range of $\pm \sim 20\%$, the RGI V5 estimates are relatively closer to the BI estimates. The GSI inventory statistically appears to be significantly different from rest of the inventories in terms of the total number of glaciers, the total glacierised area, and the area uncertainty (Table 6.1). However, the OI is almost similar for all of the test inventories, which suggests that the issues leading to area overestimation (e.g., the marking of snow and ice patches as glaciers, and the inclusion of periglacial debris, lateral moraines, nunataks, and rock glaciers within the glacier boundaries) or underestimation (e.g., the exclusion of bodies of ice above the bergschrund that are connected to the glacier, and the exclusion of supraglacial debris cover, tributary glaciers and associated medial moraines from the glacier boundaries) are evenly distributed across the inventories and need to be discussed in detail (Figs. 6.5 and 6.6).

The GSI inventory exhibited a maximum glacierised area of ~ 228 km^2 but considerably fewer glaciers compared to the other available glacier inventories. At first glance, it may appear to be due to the older date of the GSI inventory when the GLIMS mapping guidelines were not available and also the glaciers were probably larger than what we observe today. One reason for this area overestimation was

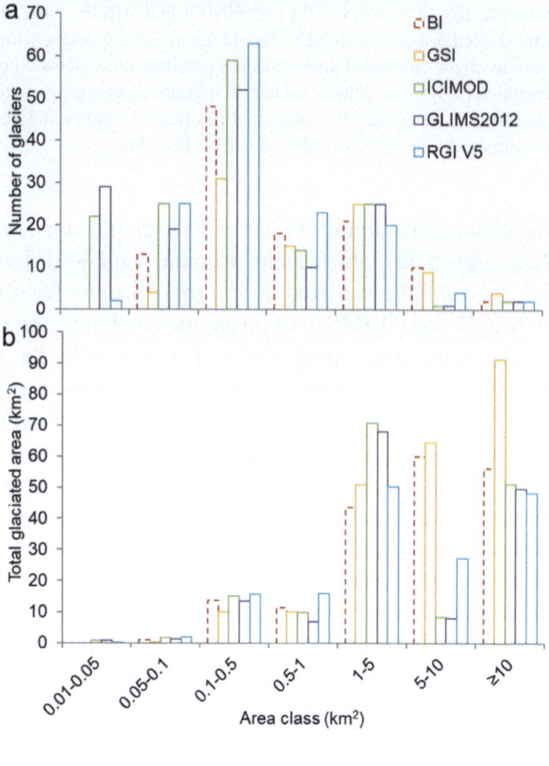

Fig. 6.4 Area class statistics of the different inventories: **a** number of glaciers, and **b** total glacierised area

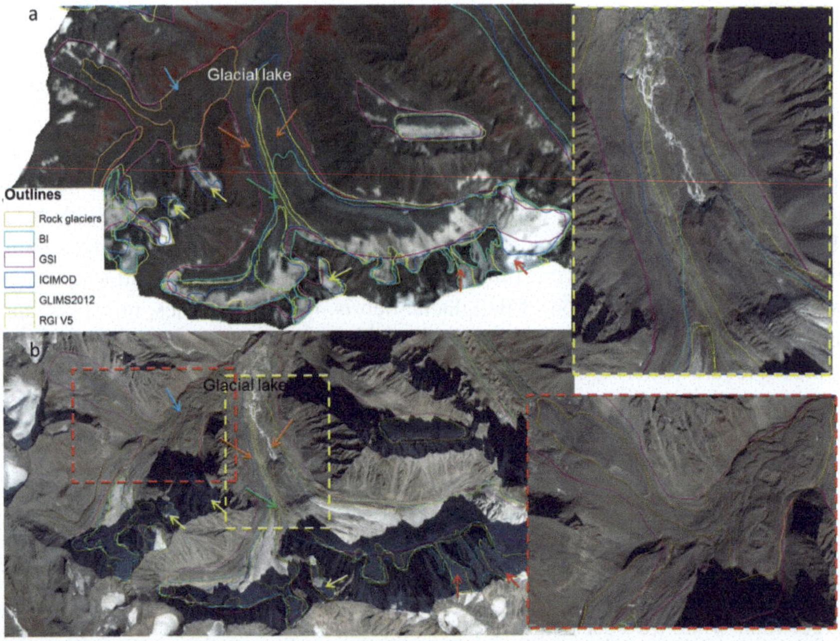

Fig. 6.5 Highlighted mapping issues for the Shaune Garang glacier. (a) 20th August 2014 Landsat 8 (scene ID: LC81460382014232LGN00) FCC (RGB: 543) image. (b) 17th December 2016 GE image. Red arrow: excluded bodies of ice above the bergschrund that are connected to the glacier; yellow arrow: included snow and ice patches; blue arrow: included rock glacier; orange arrow: included periglacial debris and lateral moraine; green arrow: excluded tributary glacier and associated medial moraine. The contextual information can be derived from Fig. 6.2. The data provider for the used Google Earth images is CNES/Airbus

the characterization of 35 out of 73 rock glaciers/ice-debris complexes as glaciers (e.g., Fig. 6.5b) which alone accounted for ~ 10 km^2 of the total area. The RGI V5 successfully excluded all the rock glaciers from the glacier outlines while the ICIMOD and GLIMS2012 inventories included one of the rock glaciers within the glacier boundaries. Another prominent reason for the area overestimation in the GSI inventory was the inclusion of periglacial debris cover and lateral moraines into the area of several heavily debris-covered glaciers (e.g., Fig. 6.5a). Although the presence of nunataks was not very prominent in the study area (i.e., they were observed for only 6 glaciers), the GSI inventory failed to exclude any of the nunataks, thereby further increasing the total area. The GSI inventory also excluded the ice bodies above the bergschrunds that were connected to the glacier from the glacier boundary which is now included within the boundary as per the GLIMS guidelines (Raup and Khalsa 2007). On several instances, the GSI inventory also displayed an exclusion of supraglacial debris cover, though not as prominently as the other test inventories. However, unlike the other test inventories, the GSI inventory efficiently avoided the characterisation of snow and ice patches that were disconnected from the main

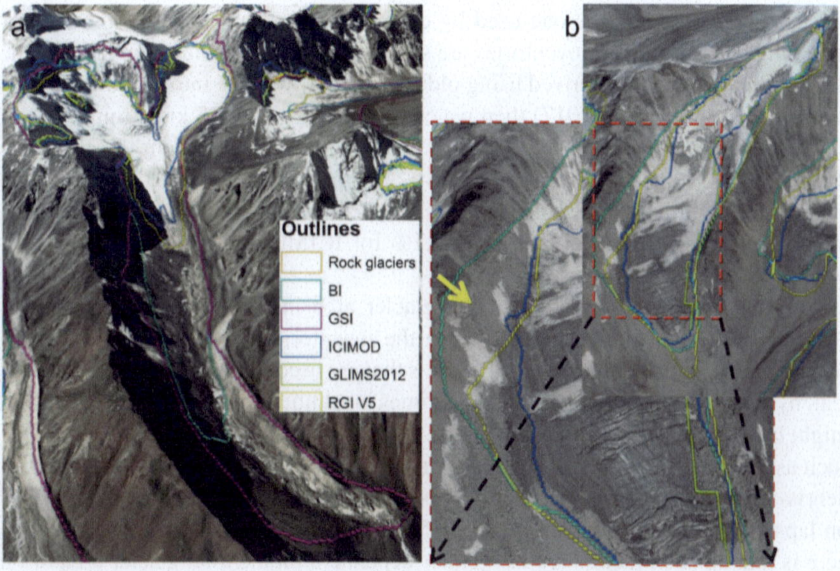

Fig. 6.6 Highlighted mapping issues in the GE images of **a** Naradu glacier (17th December 2016) and **b** Sushang glacier (25th August 2014). The yellow arrow shows the debris-covered adjoining ice excluded by the inventories from the glacier boundaries. The legends and the contextual information can be derived from Fig. 6.2. The data provider for the used Google Earth images is CNES/Airbus

glacier body as separate glaciers. This was a major issue with the other test inventories (primarily the ICIMOD inventory) (yellow arrows in Fig. 6.5), which consequently displayed a total number of glaciers that was about one-fourth more of the number derived from the BI (Table 6.1). Nevertheless, despite the greater number of glaciers in these inventories (i.e., the ICIMOD, GLIMS2012, and RGI V5 inventories), the cumulative area values of the glacierised regions were significantly lower relative to the BI (Table 6.1). In fact, they reported areas up to ~ 14–20% lesser than the BI.

If we consider the fact that the BI was prepared using the 2014 Landsat image unlike the rest of the digital inventories (except GSI) which represent glaciers in the late 1990s and early 2000s, and account for the decadal glacier shrinkage in the basin, this area underestimation could even reach ~ 30%, comparable to the area differences reported by Paul et al. (2013) for the digitisation of the same glaciers by different analysts. To be more precise, we use the published data (Mir et al. 2017) on the recession of Baspa Basin glaciers reporting an average rate of glacier area loss in this region to be ~ 1 km^2 per annum. The RGI and GLIMS outlines were derived using 2001 images while the ICIMOD outlines used 2008 images for our study area. Now, Mir et al.'s (2017) estimate is also based on Landsat remote sensing observations and cannot be taken as an absolute measure of area loss as we have seen the probable misestimates on the debris-covered parts while using the Landsat resolution data. Nevertheless, even if we consider ~ 1 km^2 per annum as a baseline rate of glacier loss in this region, then all the used images by different inventories (except GSI) including the one used by us have been acquired within a ~ 13-year time window. Also, these inventories are showing lesser area when compared to the BI even though they are derived using older images. If we take into account the area recession rate (Mir et al. 2017) then we can further deduct ~ 13 km^2 from the RGI and GLIMS inventories (based on 2001 images) and ~ 6 km^2 from the ICIMOD (based on 2008 images) inventory to give an approximate estimate for 2014 (the year of used BI image). This makes the area uncertainties shown in Table 6.1 even more significant (18.3%, 27.2%, and 21.1% for ICIMOD, GLIMS2012, and RGI V5, respectively).

Such an underrepresentation of the glacier area in these inventories is worth considering, primarily because these are the inventories that serve as inputs into several glacio-hydro-climatological models that are used to predict future glaciological, hydrological, and meteorological regimes in addition to global sea levels. There might be several factors which can define the run-off estimations within the models, such as, precipitation on the glacierised area, differential energy balances over non-debris- and debris-covered parts, and hypsometric temperature extrapolations based on lapse rates (e.g., Snehmani et al. 2015). However, an important point to notice here is that all such factors are also either dependent on the total glacier area or on the debris and non-debris area distributions.

As visible in Fig. 6.5, for our study area, the RGI V5 inventory was effective to some extent at including the ice bodies above the bergschrunds that were connected to the glacier, even though the RGI V5 inventory includes the GGI outlines for this region, which are known to exclude thin ice on headwalls (RGI Consortium, 2015) contrary to the GLIMS guidelines (Raup and Khalsa 2007). However, the inclusion and exclusion of supraglacial and periglacial debris cover, respectively, were highly variable in the RGI V5, ICIMOD, and GLIMS2012 inventories. For example, in Fig. 6.5, each of these inventories included the periglacial debris within the glacier boundaries (orange arrow). In Fig. 6.6, however, these inventories excluded nearly all of the supraglacial debris cover for Naradu Glacier and identified only visible ice and snow as part of the glacier. The region highlighted by the green arrow in Fig. 6.5 depicts the exclusion of a tributary glacier and an associated medial moraine from the main body of the Shaune Garang glacier in the RGI V5 and ICIMOD inventories. In several instances, another error source in each of these inventories was the identification and mapping of the glacier headwalls or bergschrunds. In the areas where a bergschrund was covered with debris falling from the higher slopes (e.g., the yellow arrow in the red dashed rectangle in Fig. 6.6b of Sushang Glacier), these inventories outlined the boundary over the visible ice leaving out the debris-covered headwall and adjoining ice understandably due to the limitations posed by the used moderate resolution remote sensing data.

As visible in Fig. 6.4, the ICIMOD and GLIMS2012 inventories report 22 and 29 glaciers, respectively in the smallest area class of 0.01 - 0.05 km^2 as opposed to 0 and 2 glaciers by the GSI inventory and RGI V5, respectively. Meanwhile, the glaciers in this class within the BI are zero. This confirms the inclusion of small ice and snow patches within the GLIMS2012 and ICIMOD inventories, which are not completely manual but utilize a combination of object-oriented and pixel-based classification algorithms and manual corrections for debris-covered glaciers. For the area classes 0.05–0.1 and 0.1–0.5 km^2, the RGI V5 displays similar numbers and areas as the ICIMOD and GLIMS2012 inventories. In the next area class of 0.5–1 km^2, the RGI V5 reports considerably larger number and area of glaciers (nearly twice) than ICIMOD and GLIMS2012 inventories but closest to the BI estimates. These size ranges typically represent high elevation and small debris-covered glaciers in the basin. Even in the moderate size range of 1–5 km^2, although we observe a consistency among all of the inventories, the RGI V5 estimates are very close to the BI estimates. Significant area underestimations by the ICIMOD, and GLIMS2012 inventories are shown in the case of the larger glaciers (5–10 km^2), while the GSI and RGI V5 inventories provide closer estimates to the BI in terms of the boundary conditions as they mapped the debris-covered lower reaches of the larger glaciers better. These data highlight the fact that the ICIMOD and GLIMS2012 inventories fail to accurately map the supraglacial debris cover of large-sized glaciers. For two of the largest glaciers ($\geq$ 10 km^2) in the river basin which covered more than one-fourth of the total glacierised area and clearly displayed the moraine boundaries and the huge snouts, there was a consistency among all the inventories (except GSI).

6.3.2 *Hypsometric Inconsistencies*

We also investigated the area-elevation distribution (hypsometry) (Strahler 1952) derived for these inventories (Fig. 6.7). Glacier hypsometry is useful in understanding the long-term glacier response through its correlation with mass transfer (Furbish and Andrews 1984), mass-balance elevation distribution (Brozovic et al. 1997; Jiskoot et al. 2009; Small 1995), paleotectonic (Montgomery et al. 2001), meteorological extrapolations (Snehmani et al. 2015), glacio-hydrological modelling (Kumar et al. 2016), and glacier erosion and landform evolution (Brocklehurst and Whipple 2004; Pedersen 2010; Sternai et al. 2011; Strahler 1952). Such studies at a mountain range-scale involve the use of regional or global glacier inventories. In Fig. 6.3, we present the hypsometric curves derived for a given DEM (here, the ASTER GDEM V2) classified within 13 elevation zones (each of 200 m) using each of the inventories for the glaciers of the Baspa River Basin.

The absolute uncertainty shown in Fig. 6.7 is calculated by subtracting the area estimated using the freely available inventory from the area calculated for the BI for a given elevation zone. For the lower and middle elevations, the GSI inventory shows extensive area overestimations, partly due to larger glacier coverage in the past and partly due to the significant inclusion of periglacial landforms within the glacier boundaries. In the same elevations, the RGI V5, ICIMOD, and GLIMS2012 inventories show area underestimations (Fig. 6.7). However, for the elevations higher than ~ 5,200 m, the RGI V5 hypsometric curve is in very good agreement with the BI curve. Each of the inventories provide consistent area estimates in the higher elevation zones (>5,600 m), as these elevation zones are prominently debris-free and are easier to delineate. In > 5,000 m elevation zones, the RGI V5 gives the most consistent

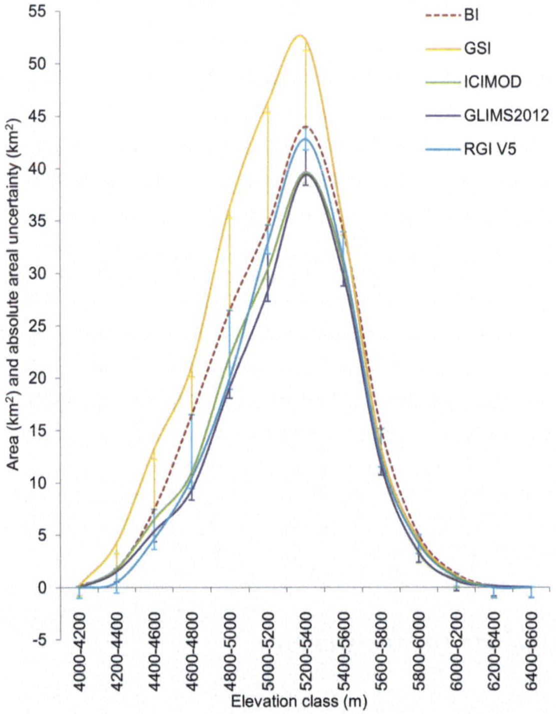

Fig. 6.7 Hypsometric profiles for the different Baspa River basin glacier inventories. The directions and the magnitudes of the error bars represent a positive (upwards) or negative (downwards) absolute uncertainty in the area value within a particular elevation zone of each of the test inventory when compared to the BI. The area-elevation information was extracted from the ASTER GDEM V2

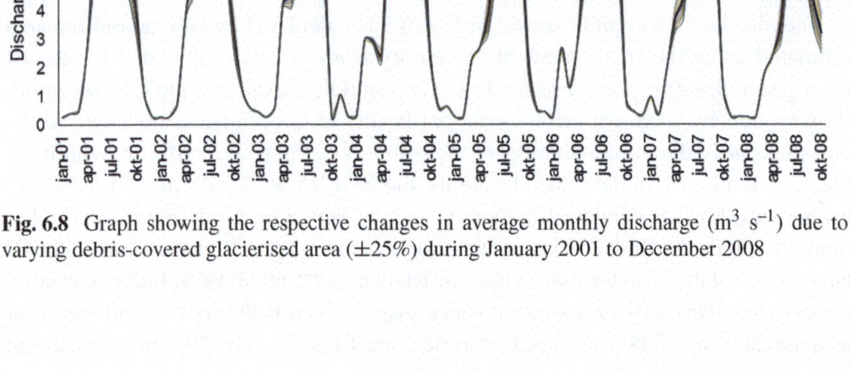

Fig. 6.8 Graph showing the respective changes in average monthly discharge ($m^3\ s^{-1}$) due to varying debris-covered glacierised area (±25%) during January 2001 to December 2008

hypsometric estimates, which is evident from its smallest absolute uncertainty bars in Fig. 6.7. However, the significant area uncertainties in the zones of supraglacial debris cover in each of these inventories is particularly alarming with respect to performing glacio-hydro-climatological modelling in these mountains heavily exhibiting debris-covered glaciers. The presence of debris cover (and the associated albedo) further introduces several uncertainties into the modelling (Kumar et al. 2016), and these issues are aggravated if the extent of the debris cover is misestimated with such large margins of uncertainties.

6.3.3 Glacio-Hydrological Inconsistencies

As mentioned in the previous section, we identified high areal uncertainty in the mapped glacierised land cover, particularly in the supraglacial debris cover. To assess the respective magnitude of changes in glacio-hydrological model output due to uncertainty in debris-covered glacierised area (−25% to + 25%), we applied the model in Shaune Garang Catchment of Baspa River Basin where it was also validated against observed data for the part of ablation season of 2014 ($R^2 = 0.89$; Kumar et al. 2016) and 2015 ($R^2 = 0.91$; Singh et al. 2018). About 24% of the total glacierised area in Shaune Garang Catchment is covered by debris of varying thickness which was observed to have a negative gradient with altitude (Kumar et al. 2016).

We have observed in the present study that various inventories have predominantly underestimated the glacierised area in general and debris-covered glacierised area in particular in Baspa Basin. However, in case of Shaune Garang Glacier an overestimation of debris-covered glacierised area was observed in all the inventories. Therefore, we used the model to estimate continuous average monthly discharge from 2001 to 2008 with changing debris-covered glacierised area between ± 25% of the total glacierised area (Fig. 6.8). This analysis shows that the discharge is directly proportional to the change in debris-covered glacierised area with response observed prominently during the months of ablation season (May to August). The change in debris-covered area by ± 25% resulted in alteration of average monthly discharge by up to ± 16%. In addition to the monthly discharge, we also used the model to estimate the daily discharge during the summer of 2014 (Kumar et al. 2016) after changing the debris cover by ± 25%, in order to compare the modelled output with in-situ observed daily discharge. The modelled discharge showed a corresponding overestimation of 30% and an underestimation of around 3% in comparison to the observed daily discharge after increasing and decreasing the debris-covered area, respectively. The variation in the modelled discharge is caused by a general overestimation of around 12% by the model (Kumar et al. 2016). The prime conclusion of this analysis is that any areal uncertainty related to the input inventory is bound to facilitate proportionally high uncertainties in the reconstructed discharge. This uncertainty is even more crucial for Himalayan glaciers which are significantly debris-covered.

Research areas that involve glacio-hydrological modelling and runoff estimations at various temporal and spatial scales to predict the future water availability extensively require various inventory inputs, such as the distribution, area, length, and hypsometry of glaciers. Many of such modelling efforts were performed for the high mountains of Asia since these regions are the most limited in terms of glacio-hydrometeorology data and are difficult to access, posing serious model uncertainties due to multiple climatic and physiographic regimes (Bhardwaj et al. 2019; Sam et al. 2019) and heavily debris-covered glaciers (Pellicciotti et al. 2012; Wester et al. 2019). The rivers that originate from these mountains are the main source of fresh water for some of the most densely populated regions on earth, which warrant exceptional modelling and simulation efforts (Immerzeel et al. 2010; Pritchard 2017).

6.3.4 Implications of Area Inconsistencies for Models

Our emphasis in this paper is on the regional- or global-scale inventories whose data are readily available. The following discussion stands valid for all the studies that used inventory data but did not consider the effects of inherent errors on the modelling results. Here, we discuss several regional-scale studies concerning the Himalaya or global-scale studies with a separate emphasis on the Himalaya. The issue of misestimation of areas in inventories of debris-covered glaciers and its implications for geophysical modelling results are widespread for similar debris-covered and small-sized glaciers in the Andes, the Alps, Alaska, or elsewhere. Thus, the discussions and conclusions in the following sections can be extrapolated at a global scale.

6.3.5 Glacio-Hydrology: Studies from Tien Shan

We understand that quantifying and integrating area misestimates within the final results is difficult in most cases due to the unavailability of exact glacier outlines for comparison. But it is extremely important to consider them in a qualitative explanation for the validity of the research and the probable scenarios (e.g., Lutz et al. 2016). Using the Glacier Inventory of China (GIC) (Ding et al. 1986a) and the USSR glacier inventory (Catalogue USSR Glaciers 1970), several authors have simulated the glacio-hydrology of the Tien Shan region. These inventories are based on similar base data (e.g., aerial, satellite, and topographic maps) and mapping methods as the other inventories described in the previous sections. Baisheng et al. (2003) used information regarding the glacier coverage, area, length, and elevation for the Yili River basin in the Tien Shan to run a glacier ice-flow model that simulated climate change-induced effects on alpine glaciers and their runoff. They suggested that the glacier size was the main factor in determining its sensitivity to climatic change. In addition to the glacier size, the rate of air temperature rise was another prominent factor affecting the runoff variations (Baisheng et al. 2003). However, Aizen et al.

(2007a) later established that the change in precipitation rather than temperature was the main driving parameter for river runoff variations in the Tien Shan. The glacier area was not only a significant model input but also an evident part of the main conclusion of the study by Baisheng et al. (2003), and inclusion of any error estimates for the input glacier sizes would have further strengthened the discussion.

Aizen et al. (2007a) used detailed data from field observations (including hydrometeorology, precipitation, glaciological measurement, stream-gauge, and evaporation site data) and glacier inventories (i.e., area data) (Catalogue USSR Glaciers, 1970; Glacier Inventory of China III 1986a,b,c, 1987) to develop a relationship between the precipitation, evapotranspiration, and summer air temperatures for runoff predictions in the Tien Shan up to 2100. They used glaciological data from six unevenly distributed glaciers (five in the north and one in the south) to develop the model for the entire Tien Shan, and they predicted the 2100 river runoff to increase by 1.047 times for an average temperature rise of 3 °C and a precipitation increase of 1.2 times. Certainly, the model of Aizen et al. (2007a) was not representative of the various climate and emission scenarios of the IPCC, but it involved a complex mix of hydro-meteoro-glaciological data and a projection for several assumed scenarios. Nonetheless, the incorporation of multiple datasets and the model dependence on the observed equilibrium line altitude (ELA) instead of the direct glacier hypsometry provided the results a degree of independence from the inventory errors.

In the same year, Aizen et al. (2007b) performed another glacio-hydrological simulation and prediction study for the Tien Shan using extensive inventory inputs, including the glacier area, number, length, elevation, and hypsometry to forecast the potential impact of global and regional climate change on the glaciers and glacier runoff. For the steady state of glaciers in Tien Shan, the model of Aizen et al. (2007b) estimated the need for a 100 mm rise in precipitation to balance a possible increase of 1 °C in the air temperature at the ELA. In another hypothetical scenario involving a possible increase of ~4 °C of the mean air temperature and a precipitation increase of ~10%, Aizen et al. (2007b) predicted an upward elevation shift of ~570 m in the ELA. Under these conditions, they also predicted the number of glaciers, area coverage, glacier volume, and runoff to be 94%, 69%, 75%, and 75% of the present-day values, respectively. With respect to the results of our sensitivity analysis, a consideration to area uncertainties in the input inventory would have helped providing a higher level of discussion on these modelled results.

6.3.6 *Glacio-Hydrology: Studies from Karakoram and Nepal*

The area estimates (69% of the present-day) of Aizen et al. (2007b) are consistent with those of a more detailed study by Immerzeel et al. (2013) for the Karakoram region (~67% of the present-day) using the representative concentration pathway 8.5 (RCP8.5) scenario. However, the volumetric estimate reported by Immerzeel et al. (2013) was nearly double, i.e., a 50% decrease compared to the 25% decrease reported by Aizen et al. (2007b). This can be attributed to the significantly different modelling approaches of the two studies. Moreover, the studies represent two completely different regions adjacent to each other, with different climates, topography, and glaciation regimes. However, the larger declines reported by Immerzeel et al. (2013) can also possibly up to an extent be ascribed to their use of the ICIMOD inventory, which represents area underestimations for debris-covered and steep accumulation parts of glaciers (Table 6.1, Fig. 6.7).

While generating runoff scenarios for the twenty-first century for different extents of deglaciation (extrapolated using present-day area coverage) in the northern Caucasus, Hagg et al. (2010) acknowledged the significant misrepresentation of debris-covered glaciers in remote sensing-based inventories. Immerzeel et al. (2013) also mentioned that the decline of glaciers in the Langtang region (Nepal) is more pronounced compared to those in the Baltoro region (Karakoram) glaciers due to their smaller sizes, which indicates the importance of the accuracy of the employed glacier outlines for such models. Immerzeel et al. (2013) recognised that variations in the projected precipitation between climate models are the major source of uncertainty in future runoff predictions. Immerzeel et al. (2012) first used the glacier outlines to clip the ASTER GDEM data and then performed the calculations on the resampled DEM (100 m resolution). They assumed an elevation of 5,500 m as the threshold for increasing or decreasing the precipitation and combined the precipitation interpolation field with the temperature interpolation field to estimate the total accumulation and total melt.

If we assume a similar misrepresentation of the lower elevations or debris-covered areas by the inventory as in our study area (Table 6.1, Fig. 6.7), then the model of Immerzeel et al. (2012) has a tendency to overestimate the accumulation or total precipitation in the higher reaches. However, they justified an estimated total annual basin precipitation of 828 mm (260% higher than the estimate of 319 mm based on the interpolated observations) by comparing it with a 688 mm/y river runoff, and they explained that the positive or steady mass balance state of the Karakoram glaciers could not account for the precipitation-runoff deficit (688 mm–319 mm = 369 mm). This gives Immerzeel et al. (2012) a runoff coefficient of 83%, which is considerable; although it could have been further justified using the additional argumentation of probable underestimates of the inventory that could bring down the runoff coefficient to 83%.

6.3.7 *Glacio-Hydrology: Studies on River Basin-Scale Modelling*

In another modelling work for the glaciers in the Lhasa River basin (Himalaya), Prasch et al. (2013) coupled regional climate model (RCM) outputs with a process-oriented glacio-hydrological model based on area, hypsometric, and ice thickness information from the GIC to assess the current and possible future contributions of ice melt to river runoff in the Lhasa River basin. With comparison to several past studies (i.e., Frauenfelder and Kääb 2009; Kääb et al. 2008), which reported a total change of -21% in the glacier area from 1970 to 2000 using the GIC and multitemporal remote sensing images, Prasch et al. (2013) showed that their glacier development model simulated an acceptable decrease of 20% over a period of 30 years. Now as we observe, in the inventories and medium-resolution satellite image interpretations concerning debris-covered glaciers, there is always an inherent area uncertainty which may reach up to ≥ 20% of underestimation in the area (Table 6.1), i.e., equivalent to several decades of glacier mass loss.

Lutz et al. (2014) used an inventory (RGI V3.2, which incorporates ICIMOD outlines) input (ice cover statistics) for cryospheric-hydrological modelling. For the five major river basins of the Himalaya (the Indus, Ganges, Brahmaputra, Salween, and Mekong River basins) they ran a cryospheric-hydrological model using the latest climate model ensemble. They quantified the upstream hydrological regimes of those rivers and assessed the future water availability in their basins. As observed for our study area, the ICIMOD inventory outlines present significant area underestimates, particularly in debris-covered regions. These observations have two prominent implications for the model of Lutz et al. (2014) that is directly proportional to the first-order area estimates: (1) the underestimation of the fractional subgrid ice cover and (2) the underestimation of the daily melt from the debris-covered parts of the glaciers. This means that there is a fair possibility of a ~15–20% underestimation (Table 6.1) of the total runoff per time step in the model as glacier melt is the prominent runoff contributor in this region (Lutz et al. 2014), which provides a buffer of 1–2 decades to the previously estimated increase in runoff until 2050 by Lutz et al. (2014). Similar implications are also applicable for several other global and regional runoff modelling studies (e.g., Bliss et al. 2014; Collier et al. 2015; Li et al. 2016; Ragettli et al. 2015) where emphasising upon the probably significant area inconsistencies within the inventories and their probable consequences on the model outputs could have added another dimension while interpreting the future scenarios. The knowledge of area uncertainty can, however, make it possible to compensate an underestimation in glacier area by modulating the degree-day factors during model calibration as an example for introducing equifinality in glacio-hydrology modelling.

6.3.8 *Studies Considering Area Uncertainties in Glacio-Hydrology Modelling*

This discussion is not complete without mentioning several studies that have acknowledged the importance of considering inventory errors while discussing the outcomes. Lutz et al. (2013) used glacier area and hypsometric information from the RGI V2.0 to run a degree-day runoff model and evaluate the uncertainties in climate change projections over the Amu Darya and Syr Darya River basins found for simulations conducted in the Coupled Model Inter-comparison Project Phase 3 (CMIP3 – Meehl et al., 2007) and Phase 5 (CMIP5—Taylor et al., 2012). During the model validation for the Austrian Alps, they categorically considered the reported estimated errors in the glacier inventories by Lambrecht and Kuhn (2007) before considering their model performance as satisfactory. In the absence of a reliable base inventory for their study area of the Upper Indus River Basin to conduct comparisons and error estimates, Mukhopadhyay and Khan (2014) assumed their area inaccuracies to be 10% of the total area while using the RGI V3.0 to run a glacio-hydro-meteorological model to estimate the contributions of different genetic sources to the melt-water running into the rivers. Although this error for a sub-basin (e.g., the Baspa River Basin) of the same river basin (e.g., the Indus River Basin) can reach up to ~ 20% (Table 6.1), the considerations provided by Mukhopadhyay and Khan (2014) offered a more realistic estimate of the runoff contributors in the higher Himalayan ranges.

For the same study area of the Indus River basin (Karakoram), another study by Lutz et al. (2016) used the RGI V3.2 (incorporating ICIMOD outlines) to derive the glacier area and hypsometric distribution along with the glacier volume, and they combined these derivatives with an ensemble of statistically downscaled circulation model outputs to force a cryospheric-hydrological model to generate transient hydrological projections. Lutz et al. (2016) referenced the Nuimura et al. (2015) study where a comparison between three different inventories for the Karakoram Range resulted in an area inconsistency of up to ~ 23%. Similarly, Frey et al. (2014) compared six different methods to calculate the glacier ice volume using inventory data for the Karakoram Range and showed an inconsistency of up to ~ 68%. Lutz et al. (2016) thus acknowledged the implications of used inventory for the simulated glacier-melt.

6.3.9 *Implications for Glacier Mass, Volume, and Sea Level Modelling*

Several studies (e.g., Cazenave and Nerem 2004; Gasson et al. 2012; Moore et al. 2013) have highlighted the substantial uncertainties involved in sea level contribution estimates. In the following paragraphs, we discuss several of the prominent studies in view of their considerations towards inventory inaccuracies. However, it is worth mentioning here that sea level projections involve a more complex methodology than glacio-hydrological modelling, and the uncertainties provided by different methods or input datasets are not directly comparable or quantifiable (Hock et al., 2009). Nevertheless, one example of the significant influence and control of the glacier area on the final modelling outcome of such a study can be found in Berthier et al. (2010) where their higher spatial resolution glacier inventory resulted in a reduction of 34% in the contribution of the Alaskan glaciers to sea levels related to previous estimates.

Wide-scale glacier volumes and mass estimates are generally derived using area information and one such approach that has been prominently presented in the literature is known as volume-area (V-A) scaling (Bahr et al. 2015). Bahr et al. (2015)

efficiently characterised some of the advantages and limitations of such scaling techniques and mentioned several published studies (e.g., Adhikari and Marshall 2012; Bliss et al. 2013; Farinotti et al. 2009; Kulkarni & Karyakarte 2014; Leclercq et al. 2011; Möller and Schneider 2010) that contained analyses that were in direct conflict (i.e., mostly arising due to an incorrect assumption of V-A scaling that it must be limited to steady state conditions and its application at an individual glacier scale) with the original V-A scaling theory by Bahr et al. (1997). The uncertainties of such estimates are further increased multifold if the area information used in the scaling process contains significant errors. Another empirical approach adopted by several studies (e.g., Chaturvedi et al., 2014; Kulkarni 1992; Kulkarni et al. 2004) to generate data for Himalayan glacier mass balances using inventory and area information is based on the debatable use of ELA and the accumulation area ratio (AAR) method.

During the initial years of this methodology, Braithwaite (1984) performed a preliminary assessment and suggested several crucial considerations while attempting this approach and interpreting the results. He concluded that, although the existing mass-balance series could be usefully extrapolated by using ELA data from additional years for a particular glacier, its generalisation and regressive extrapolation of the glaciers could give extremely erroneous and misleading estimates. Later, Kulkarni (1992) also mentioned that the results of AAR method should be favored for individual glaciers. Kulkarni et al. (2004) performed similar analyses for the glaciers of the Baspa River Basin and extrapolated a relationship obtained for only 2 glaciers to the 19 glaciers representing the entire basin. Although this was one of the early multi-glacier AAR-based mass balance extrapolations in the Indian Himalaya, any considerations on the accuracy of the used area estimates and glacier outlines employed in the study was missing. Due to the regional-scale extrapolations in this study, a consideration for the varying glaciological, physiographical, and climatological regimes of the Karakoram, Western and Middle Himalaya, and Eastern Himalaya would have possibly provided a better regional perspective.

The empirical extrapolation by Chaturvedi et al. (2014) did not consider the inherent errors in the remote sensing-based inventories as a source of uncertainty in their results, which otherwise is important in view of a detailed assessment presented by Frey et al. (2014), who proved that the V-A scaling approaches fail considerably in the context of the Himalaya, primarily because of higher inventory uncertainties. Frey et al. (2014) compared various glacier volume estimation methods using inventory data for the Himalaya, acknowledged the inventory inconsistencies, and tried to quantify the errors amounting to an estimated uncertainty of $\pm$ 4–5% in the estimates. The larger issue in the V-A scaling method arises due to the errors in delineating multi-basin glaciers with several tributaries (e.g., Baltoro Glacier). When all these tributaries do not actively contribute to the flow, they should be separated from the main glacier.

There are several prominent studies that have mentioned the probable implications of inventory inaccuracies on the modelling results pertaining to mass, volume, and sea level rise. Slangen et al. (2012) used the present-day area and volume of all of the glaciers derived from WGI-XF (Cogley 2009; Radić and Hock 2010) to evaluate the estimated sea level rise at a regional scale. They identified the possible uses of highly variable projections for different glaciological data inputs but defended their approach by claiming that it was relatively insensitive to the choice of the glacier dataset. Slangen et al. (2012) used modified WGI-XF data from Radić and Hock (2010), who had employed glacier area and distribution information from the inventory, to derive regional and global ice volumes using V-A scaling while considering the area inaccuracies, which they assumed to be 10% for each glacier, and further propagated them into the volume estimates. Leclercq et al. (2011) meticulously identified the ice coverage distribution as the main source of uncertainty in the model used in their study and provided a final uncertainty range of 25%, which is comparable with our estimates presented in Table 6.1.

6.4 Conclusions and Recommendations

This chapter investigates an otherwise little discussed or infrequently assessed aspect of cryosphere research, i.e., the implications of glacier inventory uncertainties on modelling outputs. Given below are the main conclusions and recommendations:

1. We highlight the usefulness of high-resolution and freely available GE images for modifying the glacier outlines. Considering the strong control of the ice coverage or glacier area on the models, such an uncertainty estimation will provide more reliable modelling outputs. However, GE images should only be used for a visual cross comparison. Direct correction of polygons in GE is not recommended due to the poor geometric quality of GE images.
2. We also report the highest area uncertainties for the debris-covered parts of the glaciers in the inventories. The presence of debris cover induces several unknowns, such as differential rates of melting and glacier motion.
3. The available inventories are intended to work at global or regional scales and their direct use at catchment or basin scale, especially in the high mountains, should be discouraged. However, if need arises, we recommend using RGI (please note that it is a continuously evolving compilation of various data sources for different regions) with a quick GE-based and area specific modifications for at least the Western Himalayan glacierized catchments as the RGI V5 outlines in this region most closely followed the BI outlines for majority of the glaciers, particularly in the uppermost accumulation, and lower elevation debris-covered regions.
4. In the present study we identified and mapped 73 rock glaciers/ice-debris complexes/ protalus lobes covering a wide area of ~15.17 km^2 (>8% of total glacier area). This signifies the importance of including mountain permafrost contribution to future glacio-hydrological modelling efforts in Himalayas. Considering the glacio-hydrological importance of rock glaciers (Azócar and Brenning 2009; Schmid et al. 2015; Singh et al. 2016b) and their use as a proxy for the definite presence of permafrost (Bhardwaj et al. 2016c; Jakob 1992), we recommend streamlining the data compilation and mapping of rock glaciers for

a separate repository that can be temporarily updated by researchers with field data. This will also help with methodological advances in rock glacier identification and mapping and distinguishing them from completely debris-covered glaciers.

5. The generalised implications of a probable area underestimation in an input inventory for glacio-hydrological models, runoff estimations, and water availability can be viewed in terms of providing a temporal buffer of several additional years/ decades of runoff increases and water availability compared with previous estimates. Similarly, the general implications and indications for glacier mass and volume estimates can be shifted towards higher values relative to those previously estimated in the majority of the studies. It also suggests a greater contribution to sea level rise if the present rate of global warming persists. However, we must not overlook the complexities of modelling while making these generalisations, as area miscalculates are more pronounced for debris-covered regions, which behave differentially in response to melting depending on the thickness of the debris cover (e.g., Pratap et al. 2015).
6. Sentinel-2 and PlanetScope images are free-of-cost and of considerably higher temporal and spatial resolutions in visible spectrum and can act as useful input sources for future inventory generations or updates.

Finally, we support the idea proposed by Braithwaite and Raper (2007) regarding the compilation of a complete, precise, and 'useful minimum' inventory consisting of information on single glacier elevations (e.g., mean or median elevation) and a total but accurate area. Such a compilation will take lesser time than the compilation of conventional multi-parameter inventories and will provide analysts with the ability to focus more on ensuring the quality of area accuracies which are a major source of misestimates in glacio-hydrological and sea level models. The temporal updating of such an inventory can provide a long-lasting database for assessing our cryosphere resources in many years to come. The suggestions proposed by Salvatore et al. (2015) can be helpful in this regard, where acquisition of repeated aerial photos and keeping them open access for the scientific community can contribute significantly in temporal updating of the existing inventories. The advent of unmanned aerial vehicles (UAVs) in glaciology (Bhardwaj et al. 2016b; Gaffey and Bhardwaj 2020; Gaffey et al. 2022) is revolutionary and has brought down the aerial data acquisition costs to unprecedented minimal levels. There is an emerging need for improving our assessment of present and future water availability, to effectively facilitate climate change adaptation, sustainable development, community-based practices, and risk communication in high-mountains. The results of this work will further guide glacio-hydrology modelling practices, on how to link them better to theory, data, and evidence.

Acknowledgements We extend our thanks to the Contribution to High Asia Runoff from Ice and Snow (CHARIS) project, funded by the United States Agency for International Development (USAID) for financing the glacio-hydrological field work in Baspa River Basin. The present study is not a direct derivative or objective of the CHARIS project but the field work during this project increased our familiarity with the study area and we were able to take field observations and photographs for the BI.

References

ASTER GDEM: A product of the Ministry of Economy, Trade, and Industry (METI) and the National Aeronautics and Space Administration (NASA). (downloaded from https://earthexplorer.usgs.gov Adhikari, S., and S. J. Marshall (2012), Glacier volume-area relation for high-order mechanics and transient glacier states, Geophysical Research Letters, L16505, Doi: https://doi.org/10.1029/2012GL052712.

Adhikari S, Marshall SJ (2012) Glacier volume-area relation for high-order mechanics and transient glacier states. Geophys Res Lett L16505. https://doi.org/10.1029/2012GL052712

Aizen VB, Aizen EM, Kuzmichenok VA (2007a) Geo-informational simulation of possible changes in Central Asian water resources. Global Planet Change 56(3):341–358

Aizen VB, Aizen EM, Kuzmichonok VA (2007b) Glaciers and hydrological changes in the Tien Shan: simulation and prediction. Environ Res Lett 2(4):045019

Arendt AA, Echelmeyer KA, Harrison WD, Lingle CS, Valentine VB (2002) Rapid wastage of Alaska glaciers and their contribution to rising sea level. Science 297(5580):382–386

Arora M, Rathore DS, Singh RD, Kumar R, Kumar A (2010) Estimation of Melt Contribution to Total Streamflow in River Bhagirathi and River Dhauli Ganga at Loharinag Pala and Tapovan Vishnugad Project Sites. J Water Resour Prot 2:636–643. https://doi.org/10.4236/jwarp.2010.27073

Azam MF, Wagnon P, Vincent C, Ramanathan AL, Linda A, Singh VB (2014a) Reconstruction of the annual mass balance of Chhota Shigri glacier, Western Himalaya, India, since 1969. Ann Glaciol 55(66):69–80. https://doi.org/10.3189/2014AoG66A104

Azam MF, Wagnon P, Vincent C, Ramanathan AL, Favier V, Mandal A and Pottakkal JG (2014b), Processes governing the mass balance of Chhota Shigri Glacier (western Himalaya, India) assessed by point-scale surface energy balance measurements. The Cryosphere 27;8(6), 2195–217, Doi: https://doi.org/10.5194/tc-8-2195-2014

Azócar GF, Brenning A (2009) Hydrological and geomorphological significance of rock glaciers in the dry Andes, Chile (27°-33°S). Permafrost Periglac Process 21:42–53. https://doi.org/10.1002/ppp.669

Bahr, D. B., M. F. Meier and S. D. Peckham (1997) The physical basis of glacier volume-area scaling, Journal of Geophysical Research, 102(B9), 20, 355–362, Doi: https://doi.org/10.1029/97JB01696.

Bahr DB, Pfeffer WT, Kaser G (2015) A review of volume-area scaling of glaciers. Rev Geophys 53(1):95–140

Baisheng Y, Yongjian D, Fengjing L, Caohai L (2003) Responses of various-sized alpine glaciers and runoff to climatic change. J Glaciol 49(164):1–7

Bajracharya, S. R. and B. Shrestha (2011) The status of glaciers in the Hindu Kush-Himalayan region. Kathmandu: ICIMOD, (http://lib.icimod.org/record/9419).

Barry RG (2006) The status of research on glaciers and global glacier recession: a review. Prog Phys Geogr 30(3):285–306

Bedford D, Haggerty C (1996) New digitized glacier inventory for the former Soviet Union and China. Earth Syst Monit 6(3):8–10

Berthier E, Schiefer E, Clarke GK, Menounos B, Rémy F (2010) Contribution of Alaskan glaciers to sea-level rise derived from satellite imagery. Nat Geosci 3(2):92–95

Bhardwaj, A., L. Sam and F. J. Martín-Torres (2016c) Rock glaciers as proxies for identifying terrestrial and analogous Martian permafrost. Günther, F. and A. Morgenstern (Eds.) (2016): XI. International Conference on Permafrost - Book of Abstracts, 20–24 June 2016, Potsdam, Germany. Bibliothek Wissenschaftspark Albert Einstein. pp 535–537. doi:https://doi.org/10.2312/GFZ.LIS.2016.001.

Bhardwaj A, Sam L, Akanksha FJ, Martín-Torres, and R. Kumar, (2016b) UAVs as remote sensing platform in glaciology: Present applications and future prospects. Remote Sens Environ 175:196–204

Bhardwaj A, Sam L, Singh S, Kumar R (2016c) Automated detection and temporal monitoring of crevasses using remote sensing and their implications in glacier dynamics. Ann Glaciol 57(71):81–91. https://doi.org/10.3189/2016AoG71A496

Bhardwaj A, Singh MK, Joshi PK, Singh S, Sam L, Gupta RD, Kumar R (2015a) A lake detection algorithm (LDA) using Landsat 8 data: A comparative approach in glacial environment. Int J Appl Earth Obs Geoinf 38:150–163

Bhardwaj A, Joshi PK, Sam L, Snehmani. (2016d) Remote sensing of alpine glaciers in visible and infrared wavelengths: a survey of advances and prospects. Geocarto Int 31(5):557–574

Bhardwaj A, Joshi PK, Snehmani L, Sam MK, Singh S, Singh RK (2015b) Applicability of Landsat 8 data for characterising glacier facies and supraglacial debris. Int J Appl Earth Obs Geoinf 38:51–64. https://doi.org/10.1016/j.jag.2014.12.011

Bhardwaj A, Joshi PK, Snehmani MK, Singh LS, Gupta RD (2014) Mapping debris-covered glaciers and identifying factors affecting the accuracy. Cold Reg Sci Technol 106–107:161–174. https://doi.org/10.1016/j.coldregions.2014.07.006

Bhardwaj, A., Kumar R., Sam, L. (2019). Analysing geospatial techniques for land degradation studies in Hindu Kush-Himalaya. A. Saikia and P. Thapa (eds.). Environmental Changes in the Himalayan Region. Springer International Publishing. DOI: https://doi.org/10.1007/978-3-030-03362-0_6

Bliss A, Hock R, Cogley JG (2013) A new inventory of mountain glaciers and ice caps for the Antarctic periphery. Ann Glaciol 54(63):191–199. https://doi.org/10.3189/2013AoG63A377

Bliss A, Hock R, Radić V (2014) Global response of glacier runoff to twenty-first century climate change. J Geophys Res Earth Surf 119(4):717–730

Braithwaite RJ (1984) Short Notes: Can the Mass Balance of a Glacier be Estimated from its Equilibrium-Line Altitude? J Glaciol 30(106):364–368

Braithwaite RJ, Raper SC (2007) Glaciological conditions in seven contrasting regions estimated with the degree-day model. Ann Glaciol 46(1):297–302

Brocklehurst S, Whipple KX (2004) Hypsometry of glacierised landscapes. Earth Surf. Processes Landforms 29(7):907–926. https://doi.org/10.1002/esp.1083

Brozovic N, Burbank DW, Meigs AJ (1997) Climatic limits on landscape development in the northwestern Himalaya. Science 276:571–574

Catalogue USSR Glaciers, 1970–1978, Middle Asia and Kazakhstan. Leningrad. Hydrometeo Publisher (Russia).

Cazenave, A., and R. S. Nerem (2004) Present-day sea level change: Observations and causes, Reviews of Geophysics, 42(3).

Chaturvedi RK, Kulkarni AV, Karyakarte Y, Joshi J, Bala G (2014) Glacial mass balance changes in the Karakoram and Himalaya based on CMIP5 multi-model climate projections. Clim Change 123(2):315–328

Cogley JG (2009) A more complete version of the World Glacier Inventory. Ann Glaciol 50(53):32–38. https://doi.org/10.3189/172756410790595859

Collier E, Maussion F, Nicholson LI, Mölg T, Immerzeel WW, Bush ABG (2015) Impact of debris cover on glacier ablation and atmosphere–glacier feedbacks in the Karakoram. Cryosphere 9(4):1617–1632

Farinotti D, Huss M, Bauder A, Funk M (2009) An estimate of the glacier ice volume in the Swiss Alps. Global Planet Change 68(3):225–231

Frauenfelder, R. and A. Kääb (2009) Glacier mapping from multitemporal optical remote sensing data within the Brahmaputra river basin, in: Proceedings of the 33rd International Symposium on Remote Sensing of Environment, Stresa, Italy, 4–8 May 2009, available at: http://folk.uio.no/kaeaeb/publications/299 R.Frauenfelder.pdf.

Frey H, Machguth H, Huss M, Huggel C, Bajracharya S, Bolch T, Kulkarni AV, Linsbauer A, Salzmann N, Stoffel M (2014) Estimating the volume of glaciers in the Himalayan-Karakoram region using different methods. Cryosphere 8(6):2313–2333

Fujita K, Ageta Y (2000) Effect of summer accumulation on glacier mass balance on the Tibetan Plateau revealed by mass-balance model. J Glaciol 46(153):244–252. https://doi.org/10.3189/172756500781832945

Fujita K, Sakai A (2014) Modelling runoff from a Himalayan debris-covered glacier. Hydrol Earth Syst Sci 18(7):2679–2694. https://doi.org/10.5194/hess-18-2679-2014

Furbish DJ, Andrews JT (1984) The use of hypsometry to indicate long-term stability and response of valley glaciers to changes in mass transfer. J Glaciol 30(105):199–211

Gaffey C, Bhardwaj A (2020) Applications of unmanned aerial vehicles in cryosphere: Latest advances and prospects. Remote Sensing 12(6):948

Gaffey CB, Bhardwaj A, Frey KE, Estes L (2022) Polar and Cryospheric Remote Sensing Using sUAS. SUAS Applications in Geography. Springer International Publishing, Cham, pp 235–261

Gardner A, Moholdt G, Cogley JG, Wouters B, Arendt A, Wahr J, Berthier E, Hock R, Pfeffer WT, Kaser G, Ligtenberg SRM, Bolch T, Martin J, Sharp MJ, Hagen JO, van den Broeke MR, Paul F (2013) A Reconciled Estimate of Glacier Contributions to Sea Level Rise: 2003 to 2009. Science 340:852–857. https://doi.org/10.1126/science.1234532

Gasson, E., M. Siddall, D. J. Lunt, O. J. Rackham, C. H. Lear, and D. Pollard (2012) Exploring uncertainties in the relationship between temperature, ice volume, and sea level over the past 50 million years, Reviews of Geophysics, 50(1), RG1005, 1–35.

GGI: Nuimura et al. (2015) Provided by Dr. Akiko Sakai (shakai@nagoya-u.jp), Principal Investigator of the GAMDAM project http://www.cryoscience.net/pages/gamdam.html

GSI: Raina and Srivastava (2008)

GLIMS: GLIMS, and the National Snow and Ice Data Center [2005, updated 2012] (http://www.glims.org/maps/glims)

Glacier Inventory of China III (1986a) Tianshan Mountains (Ili Drainage Basin). China Science Press, Lanzhou Institute of Glaciology and Geocryology

Glacier Inventory of China III (1986b) Tianshan Mountains (Interior Drainage Area of Scattered Flow in East). China Science Press, Lanzhou Institute of Glaciology and Geocryology

Glacier Inventory of China III (1986c) Tianshan Mountains (Interior Drainage Area of Junggar Basin, Northwest). China Science Press, Lanzhou Institute of Glaciology and Geocryology

Glacier Inventory of China III (1987) Tianshan Mountains (Interior Drainage Area of Tarim Basin, Southwest). China Science Press, Lanzhou Institute of Glaciology and Geocryology

GLIMS Consortium (2005). GLIMS Glacier Database, Version 1 [Data Set]. Boulder, Colorado USA. NASA National Snow and Ice Data Center Distributed Active Archive Center. https://doi.org/10.7265/N5V98602

Guo W, Liu S, Xu SJ, Wu L, Shangguan D, Yao X, Wei J, Bao W, Yu P, Liu Q, Jiang Z (2015) The second Chinese glacier inventory: data, methods and results. J Glaciol 61:357–371. https://doi.org/10.3189/2015JoG14J209

Hagg W, Hoelzle M, Wagner S, Mayr E, Klose Z (2013) Glacier and runoff changes in the Rukhk catchment, upper Amu-Darya basin until 2050. Global Planet Change 110:62–73

Hagg W, Shahgedanova M, Mayer C, Lambrecht A, Popovnin V (2010) A sensitivity study for water availability in the Northern Caucasus based on climate projections. Global Planet Change 73(3):161–171

Hirabayashi Y, Zang Y, Watanabe S, Koirala S, Kanae S (2013) Projection of glacier mass changes under a high-emission climate scenario using the global glacier model HYOGA2. Hydrological Research Letters 7(1):6–11

Hock R (2003) Temperature index melt modelling in mountain areas. J Hydrol 282(1–4):104–115. https://doi.org/10.1016/S0022-1694(03)00257-9

Hock, R., M. de Woul, V. Radić, and M. Dyurgerov (2009) Mountain glaciers and ice caps around Antarctica make a large sea-level rise contribution, Geophysical Research Letters, 36(7), L07501, 1–5.

Immerzeel WW, Pellicciotti F, Shrestha AB (2012) Glaciers as a proxy to quantify the spatial distribution of precipitation in the Hunza basin. Mt Res Dev 32(1):30–38

ICIMOD: Bajracharya and Shrestha [2011] (http://rds.icimod.org/Home/DataDetail?metadataId=9361) (http://rds.icimod.org/dataviewer/index.html?url=http://apps.geoportal.icimod.org/ArcGIS/rest/services/Cryosphere/Glacier/MapServer/0&id=9361)

Immerzeel WW, Pellicciotti F, Bierkens MFP (2013) Rising river flows throughout the twenty-first century in two Himalayan glacierized watersheds. Nat Geosci 6(9):742–745

Immerzeel WW, Van Beek LP, Bierkens MFP (2010) Climate change will affect the Asian water towers. Science 328:1382–1385. https://doi.org/10.1126/science.1183188

Jakob M (1992) Active rock glaciers and the lower limit of discontinuous alpine permafrost. Khumbu Himalaya, Nepal, Permafrost and Periglacial Processes 3:253–256

Jiskoot H, Curran CJ, Tessler DL, Shenton LR (2009) Changes in Clemenceau Icefield and Chaba Group glaciers, Canada, related to hypsometry, tributary detachment, length–slope and area–aspect relations. Ann Glaciol 50(53):133–143

Kääb, A., R. Frauenfelder and J. A. F. Kääb (2008) Glacier distribution and glacier area changes 1960s-2000 in the Brahmaputra River basin, EGU General Assembly, Vienna, Austria, 13–18 April 2008, EGU08–A–05334.

Kargel JS, Leonard GJ, Bishop MP, Kääb A, Raup BH (2014) Global Land Ice Measurements from Space. Springer. https://doi.org/10.1007/978-3-540-79818-7

Kaser G, Großhauser M, Marzeion B (2010) Contribution potential of glaciers to water availability in different climate regimes. Proceeding of National Academy of Sciences, USA 107(20223–20227):2010. https://doi.org/10.1073/pnas.1008162107

Kaul, M. K. (1999) Inventory of the Himalayan glaciers. Calcutta, Geological Survey of India. (GSI Special Publication 34).

Kayastha RB, Ohata T, Ageta Y (1999) Application of a mass-balance model to a Himalayan glacier. J Glaciol 45(151):559–567. https://doi.org/10.3189/S002214300000143X

Kulkarni AV (1992) Mass balance of Himalayan glaciers using AAR and ELA methods. J Glaciol 38(128):101–104

Kulkarni AV, Karyakarte Y (2014) Observed changes in Himalayan glaciers. Curr Sci 106(2):237–244

Kulkarni AV, Rathore BP, Alex S (2004) Monitoring of glacial mass balance in the Baspa basin using accumulation area ratio method. Curr Sci 86(1):185–190

Kumar P, Kotlarski S, Moseley C, Sieck K, Frey H, Stoffel M, Jacob D (2015) Response of Karakoram-Himalayan glaciers to climate variability and climatic change: A regional climate model assessment. Geophys Res Lett 42(6):1818–1825

Kumar R, Kumar R, Singh S, Singh A, Bhardwaj A, Kumari A, Randhawa SS, Saha A (2018) Dynamics of suspended sediment load with respect to summer discharge and temperatures in Shaune Garang glacierized catchment. Western Himalaya. Acta Geophysica 66(5):1109–1120

Kumar R, Singh S, Kumar R, Singh A, Bhardwaj A, Sam L, Randhawa SS, Gupta A (2016) Development of a Glacio-hydrological Model for Discharge and Mass Balance Reconstruction. Water Resour Manage 30(10):3475–3492

Lambrecht A, Kuhn M (2007) Glacier changes in the Austrian Alps during the last three decades, derived from the new Austrian glacier inventory. Ann Glaciol 46:177–184. https://doi.org/10.3189/172756407782871341

Leclercq PW, Oerlemans J, Cogley JG (2011) Estimating the glacier contribution to sea-level rise for the period 1800–2005. Surv Geophys 32(4–5):519–535

Lejeune Y, Bertrand JM, Wagnon P, Morin S (2013) A physically based model of the year-round surface energy and mass balance of debris-covered glaciers. J Glaciol 59(214):327–344. https://doi.org/10.3189/2013JoG12J149

Li H, Xu CY, Beldring S, Tallaksen LM, Jain SK (2016) Water resources under climate change in Himalayan basins. Water Resour Manage 30(2):843–859

Lutz AF, Immerzeel WW, Shrestha AB, Bierkens MFP (2014) Consistent increase in High Asia's runoff due to increasing glacier melt and precipitation. Nat Clim Chang 4(7):587–592

Lutz AF, Immerzeel WW, Gobiet A, Pellicciotti F, Bierkens MFP (2013) Comparison of climate change signals in CMIP3 and CMIP5 multi-model ensembles and implications for Central Asian glaciers. Hydrol Earth Syst Sci 17(9):3661–3677

Lutz AF, Immerzeel WW, Kraaijenbrink PDA, Shrestha AB, Bierkens MFP (2016) Climate Change Impacts on the Upper Indus Hydrology: Sources. Shifts and Extremes, PLoS One 11(11):e0165630

Marzeion B, Jarosch AH, Hofer M (2012) Past and future sealevel change from the surface mass balance of glaciers. Cryosphere 6:1295–1322. https://doi.org/10.5194/tc-6-1295-2012

Meehl GA, Coauthors, (2007) The WCRP CMIP3 multimodel dataset: A new era in climate change research. Bull. Amer. Meteor. Soc. 88:1383–1394. https://doi.org/10.1175/BAMS-88-9-1383

Miles ES, Pellicciotti F, Willis IC, Steiner JF, Buri P, Arnold NS (2016) Refined energy-balance modelling of a supraglacial pond, Langtang Khola. Nepal. Annals of Glaciology 57(71):29–40. https://doi.org/10.3189/2016AoG71A42

Mir RA, Jain SK, Jain SK, Thayyen RJ, Saraf AK (2017) Assessment of recent glacier changes and its controlling factors from 1976 to 2011 in Baspa basin, western Himalaya. Arct Antarct Alp Res 49(4):621–647. https://doi.org/10.1657/AAAR0015-070

Mölg T, Maussion F, Yang W, Scherer D (2012) The footprint of Asian monsoon dynamics in the mass and energy balance of a Tibetan glacier. Cryosphere 6(6):1445–1461. https://doi.org/10.5194/tc-6-1445-2012

Möller M, Schneider C (2010) Calibration of glacier volume–area relations from surface extent fluctuations and application to future glacier change. J Glaciol 56(195):33–40

Montgomery DB, Balco G, Willett SD (2001) Climate, tectonics and the morphology of the Andes. Geology 29:579–582. https://doi.org/10.1130/0091-7613(2001)029%3c0579:CTATMO%3e2.0.CO;2

Mool, P. K., D. Wangda, S. R. Bajracharya, K. Kunzang, D. R. Gurung and S. P. Joshi (2001) Inventory of Glaciers, Glacial Lakes and Glacial Lake Outburst Floods, Bhutan, International Centre for Integrated Mountain Development (ICIMOD), Kathmandu, Nepal, 227.

Moore JC, Grinsted A, Zwinger T, Jevrejeva S (2013) Semiempirical and process-based global sea level projections. Rev Geophys 51(3):484–522

Mukhopadhyay B, Khan A (2014) A quantitative assessment of the genetic sources of the hydrologic flow regimes in Upper Indus Basin and its significance in a changing climate. J Hydrol 509:549–572

Nagai H, Fujita K, Sakai A, Nuimura T, Tadono T (2016) Comparison of multiple glacier inventories with a new inventory derived from high-resolution ALOS imagery in the Bhutan Himalaya. Cryosphere 10(1):65–85

Nuimura T, Sakai A, Taniguchi K, Nagai H, Lamsal D, Tsutaki S, Kozawa A, Hoshina Y, Takenaka S, Omiya S, Tsunematsu K, Tshering P, Fujita K (2015) The GAMDAM glacier inventory: a quality-controlled inventory of Asian glaciers. Cryosphere 9(3):849–864. https://doi.org/10.5194/tc-9-849-2015

Pedersen VK (2010) Alpine glacial topography and the rate of rockcolumn uplift. Geomorphology 122(1–2):129–139. https://doi.org/10.1016/j.geomorph.2010.06.005

Pellicciotti F, Buergi C, Immerzeel WW, Konz M, Shrestha AB (2012) Challenges and uncertainties in hydrological modeling of remote Hindu Kush–Karakoram–Himalayan (HKH) basins: suggestions for calibration strategies. Mt Res Dev 32(1):39–50

Pfeffer, W. T., A. A. Arendt, A. Bliss, T. Bolch, J. G. Cogley, A. S. Gardner, J. O. Hagen, R. Hock, G. Kaser, C. Kienholz, E. S. Miles, G. M. N. Molg, F. Paul, V. Radic, P. Rastner, B. H. Raup, J. Rich, M. J. Sharp, The Randolph Consortium (2014) The Randolph Glacier Inventory: a globally complete inventory of glaciers. J Glaciol 60(221):537–552

Pradhananga NS, Kayastha RB, Bhattarai BC, Adhikari TR, Pradhan SC, Devkota LP, Shrestha AB, Mool PK (2014) Estimation of discharge from Langtang River basin, Rasuwa, Nepal, using a glacio-hydrological model. Ann Glaciol 55(66):223–230. https://doi.org/10.3189/2014AoG66A123

Prasch M, Mauser W, Weber M (2013) Quantifying present and future glacier melt-water contribution to runoff in a central Himalayan river basin. Cryosphere 7(3):889

Pratap B, Dobhal DP, Mehta M, Bhambri R (2015) Influence of debris cover and altitude on glacier surface melting: a case study on Dokriani Glacier, central Himalaya. India. Annals of Glaciology 56(70):9–16

Pritchard HD (2017) Asia's glaciers are a regionally important buffer against drought. Nature 169–174. https://doi.org/10.1038/nature22062

Racoviteanu AE, Paul F, Raup B, Khalsa SJS, Armstrong R (2009) Challenges and recommendations in mapping of glacier parameters from space: results of the 2008 Global Land Ice Measurements from Space (GLIMS) workshop, Boulder, Colorado, USA. Ann Glaciol 50(53):53–69

RGI V5.0: RGI Consortium (2015). https://doi.org/10.7265/gq4p-zz56

Radić V, Hock R (2011) Regionally differentiated contribution of mountain glaciers and ice caps to future sea-level rise. Nature Geosciences 4:91–94. https://doi.org/10.1038/ngeo1052

Radić V, Hock R (2013) Glaciers in the Earth's hydrological cycle: assessments of glacier mass and runoff changes on global and regional scales, In The Earth's Hydrological Cycle (813–837). Springer, Netherlands

Radić, V., and R. Hock (2010) Regional and global volumes of glaciers derived from statistical upscaling of glacier inventory data. Journal of Geophysical Research: Earth Surface, 115(F1).

Radok U (1997) The International Commission on Snow and Ice (ICSI) and its precursors, 1894–1994. Hydrol Sci J 42(2):131–140

Ragettli S, Pellicciotti F, Immerzeel WW, Miles ES, Petersen L, Heynen M, Shea JM, Stumm D, Joshi S, Shrestha A (2015) Unraveling the hydrology of a Himalayan catchment through integration of high resolution in situ data and remote sensing with an advanced simulation model. Adv Water Resour 78:94–111

Raina VK, Srivastava D (2008) Glacier atlas of India. Geological Society of India, Bangalore

Raup B, Khalsa SJS (2007) GLIMS analysis tutorial. Boulder, CO, University of Colorado, National Snow and Ice Data Center, available at: http://www.glims.org/MapsAndDocs/guides.html, Accessed on: 20 October 2014.

Raup B, Kääb A, Kargel JS, Bishop MP, Hamilton G, Lee E, Paul F, Rau F, Soltesz D, Khalsa SJS, Beedle M, Helm C (2007) Remote sensing and GIS technology in the Global Land Ice Measurements from Space (GLIMS) Project. Comput Geosci 33(1):104–125. https://doi.org/10.1016/j.cageo.2006.05.015

RGI Consortium (2015). Randolph Glacier Inventory - A Dataset of Global Glacier Outlines, Version 5 [Data Set]. Boulder, Colorado USA. National Snow and Ice Data Center. https://doi.org/10.7265/gq4p-zz56.

Sakai A, Nuimura T, Fujita K, Takenaka S, Nagai H, Lamsal D (2015) Climate regime of Asian glaciers revealed by GAMDAM glacier inventory. Cryosphere 9:865–880. https://doi.org/10.5194/tc-9-865-2015

Salvatore MC, Zanoner T, Baroni C, Carton A, Banchieri FA, Viani C, Giardino M, Perotti L (2015) The state of Italian glaciers: a snapshot of the 2006–2007 hydrological period. Geogr Fis Din Quat 38(2):175–198

Sam L, Kumar R, Bhardwaj A (2019) Glaciers as indicator of changing climate in Himalaya. A. Saikia and P. Thapa (eds.). Environmental Changes in the Himalayan Region. Springer International Publishing. https://doi.org/10.1007/978-3-030-03362-0_4

Sam L, Bhardwaj A, Singh S, Kumar R (2016) Remote sensing flow velocity of debris-covered glaciers using Landsat 8 data. Prog Phys Geogr 40(2):305–321. https://doi.org/10.1177/0309133315593894

Sam L, Bhardwaj A, Kumar R, Buchroithner MF, Martín-Torres FJ (2018) Heterogeneity in topographic control on velocities of Western Himalayan glaciers. Sci Rep 8(1):12843

Schmid MO, Baral P, Gruber S, Shahi S, Shrestha T, Stumm D, Wester P (2015) Assessment of permafrost distribution maps in the Hindu Kush Himalayan region using rock glaciers mapped in Google Earth. Cryosphere 9(6):2089–2099

Shea JM, Immerzeel WW, Wagnon P, Vincent C, Bajracharya S (2015) Modelling glacier change in the Everest region. Nepal Himalaya. the Cryosphere 9(3):1105–1128

Shekhar M, Bhardwaj A, Singh S, Ranhotra PS, Bhattacharyya A, A. K., Pal, I. Roy, F.J Martín-Torres and M.-P. Zorzano, (2017) Himalayan glaciers experienced significant mass loss during later phases of little ice age. Sci Rep 7(1):10305. https://doi.org/10.1038/s41598-017-09212-2

Shi Y, Liu C, Kang E (2010) The glacier inventory of China. Ann Glaciol 50:1–4. https://doi.org/10.3189/172756410790595831

Shrestha M, Koike T, Hirabayashi Y, Xue Y, Wang L, Rasul G, Ahmad B (2015) Integrated simulation of snow and glacier melt in water and energy balance-based, distributed hydrological modeling framework at Hunza River Basin of Pakistan Karakoram region. Journal of Geophysical Research: Atmospheres 120(10):4889–4919. https://doi.org/10.1002/2014JD022666

Singh P, Bengtsson L (2003) Effect of warmer climate on the depletion of snow covered area in the Satluj basin in the western Himalayan region. Hydological Sciences Journal 48(3):413–425. https://doi.org/10.1623/hysj.48.3.413.45280

Singh P, Jain SK (2003) Modelling of streamflow and its components for a large Himalayan basin with predominant snowmelt yields. Hydological Sciences Journal 48(2):257–276. https://doi.org/10.1623/hysj.48.2.257.44693

Singh P, Umesh KH, Kumar N (2008) Modelling and estimation of different components of streamflow for Gangotri Glacier basin. Himalaya. Hydrological Sciences Journal 53(2):309–322

Singh, S., A. Bhardwaj and R. Kumar (2016b) Implications of ignoring permafrost in glacio-hydrological models in high mountains. Günther, F. and Morgenstern, A. (Eds.) (2016): XI. International Conference on Permafrost - Book of Abstracts, 20–24 June 2016, Potsdam, Germany. Bibliothek Wissenschaftspark Albert Einstein. pp 698–699. https://doi.org/10.2312/GFZ.LIS.2016.001

Singh S, Kumar R, Bhardwaj A, Kumar R, Singh A (2018) Changing climate and glacio-hydrology: a case study of Shaune Garang basin, Himachal Pradesh. International Journal of Hydrology Science and Technology 8(3):258–272

Singh S, Kumar R, Bhardwaj A, Sam L, Shekhar M, Singh A, Kumar R, Gupta A (2016b) Changing climate and glacio-hydrology in Indian Himalayan Region: a review. Wiley Interdisciplinary Reviews: Climate Change 7(3):393–410

Slangen ABA, C. A., Katsman, R. S. W. Van de Wal, L. L. A. Vermeersen and R. E. M. Riva, (2012) Towards regional projections of twenty-first century sea-level change based on IPCC SRES scenarios. Clim Dyn 38(5–6):1191–1209

Small EE (1995) Hypsometric forcing of stagnant ice margins: Pleistocene valley glaciers. San Juan Mountains, Colorado, Geomorphology 14:109–121

Snehmani A, Bhardwaj MK, Singh RD, Gupta PK, Joshi and A. Ganju, (2015) Modelling the hypsometric seasonal snow cover using meteorological parameters. J Spat Sci 60(1):51–64

Sternai P, Herman F, Fox MR, Castelltort S (2011) Hypsometric analysis to identify spatially variable glacial erosion. J Geophyss Research: Earth Surface, 116(F3)

Strahler A (1952) Hypsometric (area-altitude) analysis of erosional topography. The Geographical Society of America Bulletin 63:1117–1142. https://doi.org/10.1130/0016-7606(1952)63[1117:HAAOET]2.0.CO;2

Taylor KE, Stouffer RJ, Meehl GA (2012) An overview of CMIP5 and the experiment design. Bull. Amer. Meteor. Soc. 93:485–498. https://doi.org/10.1175/BAMS-D-11-00094.1

Vohra CP (1980) Some problems of glacier inventory in the Himalayas, World Glacier Inventory - Inventaire mondial des Glaciers (Proceedings of the Riederalp Workshop, September 1978; Actes de l'Atelier de Riederalp, September 1978): IAHS-AISH Publ. no. 126. IAHS Publ 126:67–74

Wester P, Mishra A, Mukherji A, Shrestha AB (eds) (2019) The Hindu Kush Himalaya assessment: mountains, climate change, sustainability and people. Springer Nature, Dordrecht

WGMS (1989) World glacier inventory status report 1988, Edited by: W. Haeberli, H. Bösch, K. Scherler, G. Østrem, C. C. Wallén. IAHS, GEMS/UNEP, UNESCO, Switzerland

Chapter 7
Peopling of the Sagar Island in the Indian Sundarbans: A Case of Maladaptation to Climate Change

Chinmoyee Mallik and Sunando Bandyopadhyay

Abstract The inhabited regions of the Indian Sundarbans of the Bengal Delta are threatened by dual processes of sea level rise and persistent coastal erosion caused by regular tidal surges. A few islands of this region have already been completely eroded stimulating population movements. This case study seeks to bring out how a myopic understanding of the adaptation priorities have resolved the concerns only in the short run but in the long run has exacerbated the vulnerabilities turning them maladaptive. This chapter is based upon a fieldwork in the Sagar Island comprising of 240 households in 2021 to understand the implications of the two major adaptation measures undertaken in the region to combat coastal flooding and coastal erosion: embankment and resettlement of the environmental refugee population. It is evident from the study that embankments have been able to contain coastal flooding and coastal erosion only with limited success. Instead, they have interfered with the coastal processes by altering the sediment load dynamics, reduction in channel capacity, increasing the tidal amplitude, and have exacerbated the environmental crisis. The resettlement strategy has accommodated the displaced communities, but in the long run the economic as well as the environmental outcomes have clear indications of maladaptive practices. It has deteriorated the livelihoods of the relocated communities. Further, the location of the resettlement colonies being in the fresh accretion zones have interfered with the coastal processes aggravating erosion.

Keywords Sagar Island · Indian Sundarbans · Embankment · Resettlement · Climate maladaptation · Environmental refugee

---

C. Mallik (✉)
Department of Rural Studies, West Bengal State University, West Bengal Kolkata 700126, India
e-mail: chinmoyeemallik@gmail.com

S. Bandyopadhyay
Department of Geography, University of Calcutta, West Bengal Kolkata 700019, India
e-mail: sunando@live.com

A. Sarkar et al. (eds.), *Risk, Uncertainty and Maladaptation to Climate Change*, Disaster Risk Reduction, https://doi.org/10.1007/978-981-99-9474-8_7

7.1 Introduction

The inhabited regions of the Indian Sundarbans of the Bengal Delta are threatened by persistent coastal erosion caused by dual processes of sea level rise as well as regular tidal surges. Traditional method to keep away the tidal water/storm surge from inundating the settled segments has been through construction of mud embankments. Irrespective of these efforts, several segments of the Sundarbans are eroding while some islands have already been completely eroded (Dasgupta et al. 2022) stimulating population movements. The case of erosion of the Ghoramara island within the Sagar Block in the South 24 Parganas is a case in point which has reduced to one third of its area. Communities have sought to relocate from the shrinking islands either voluntarily in search of safer locations or have been 'trapped' (Black et al. 2013; Danda et al. 2019) only to be resettled by the State Government.

Climate change and environmental stress as triggers for population movement is elaborately debated during the last few decades arguing that it is a single most important strategy to reduce vulnerability and that it is increasingly accepted as a mode of adaptation to climate change (Kothari 2014; Danda et al. 2019). The IPCC 6th Assessment Report (2022) forcefully argues that all other adaptation measures with the exception of planned relocation and migration due to sea level rise are temporary solutions to coastal hazards (IPCC 2022: 478). The sessions of the Conference of Parties (COP) clearly acknowledge the magnitude of climate hazard linked human mobility and seek to achieve mainstream planned population relocations. Several countries like Fiji, Vietnam, Philippines, and USA have sought to resettle population threatened by climate related environmental stress (Edwards 2016; Entzinger and Scholten 2016; Georgetown Climate Centre 2019; Ferris and Weerasinghe 2020) as a policy intervention to reduce vulnerability with varied implications. Often 'managed retreat' is prescribed as a mode of adaptation to ecological crisis (Danda et al. 2019).

The IPCC 6th Assessment Report (2022) even has specifically focussed upon the population mobility issues of the small islands (p. 2043) to bring out the temporaryness of the short-term benefits of the structural interventions (such as seawalls, dykes, embankments, etc.) and how some island countries have already applied 'coastal setback policies to hotels' (p. 2097) to retreat from the shorelines as a mode to reduce risks of coastal flooding, sea level rise, and coastal erosion. However, that the environmental sustainability issue may be used to camouflage a deeper political and economic project is presented by Kothari (2014) in the study of how the Maldives government have sought to relocate the population from its 200 dispersed island to selected 10–12 islands. This is done in the name of climate adaptation only to restrict the provisioning of essential services to the selected 10–12 islands instead of the 200 islands dispersed across space.

The IPCC assessment reports have been increasingly careful in understanding the environmental, social, and economic implications of any form of adaptation only to be able to clearly identify the mal-adaptive practices. The reports have highlighted the issue of maladaptation to climate change as actions aiming at climate change adaptation resulting into unintended adverse impact. It might stem from a myopic understanding of the adaptation priorities which, in the short run, may resolve the concerns but in the long run may exacerbate the vulnerabilities. The recent study on the sea-wall project in Fiji has highlighted how these dykes have not only failed to keep the tidal surges at bay but have trapped water on the landward side in addition to other negative impact upon the livelihoods (Piggott-McKellar et al. 2020). How the embankments meant for flood proofing have unintended series of negative social as well as environmental effects is underlined by Prichard and Theilmans (2014) in their study of Bihar flood management. Christian-Smith et al. (2015) have clearly indicated the complexities of the critical balance between natural water availability and societal demand and how different drought management schemes have accentuated greenhouse gas emission, groundwater depletion and in turn greater dependence upon fossil fuel-based energy sources all in totality translating into maladaptation. As argued by Reckien et al. (2023), adaptation and maladaptation may be considered as two ends of a spectrum where categorization of any action into binaries is problematic. There is time specific and locality specific expressions of adaptation practices and may be a combination of desirable as well as undesirable outcomes that make the stringent categorization complex.

This chapter is based on fieldwork conducted during 2021 in the Sagar Community Development Block (CD Block) of South 24 Parganas District in West Bengal. The region marks the westernmost fringe of the Indian Sundarbans. Sagar was started to be reclaimed from 1811 (Bandyopadhyay 1997). Embanking the low-lying islands to prevent inundation from flood tides and subsequent removal of forests had been the reclamation procedure here as in the rest of the Sundarbans. Apart from depriving the area from sediment deposition, the process exposed the settlers to a unique set of natural environmental hazards—storm inundation, saline intrusion, sea level rise, coastal erosion, and channel sedimentation—that are set to exacerbate in a warming world (Bandyopadhyay, 2021).

In this work, Remote Sensing and Geographical Information System (RS-GIS) are utilized to estimate coastal erosion. The fieldwork was conducted in the erosion-prone villages to understand the long-term implications of the two most important adaptation schemes that have been practiced in the study area: (a) the embankments and (b) resettlement of environmentally stressed communities. The study seeks to look into the environmental implications of the embankments on one hand and on the other hand attempts to highlight the broad economic and ecological outcomes of resettlement of displaced communities due to coastal erosion.

7.2 Materials and Methods

This study is based upon some fundamental RS-GIS based mapping tasks followed by a fieldwork consisting of a survey of 240 households from the zones identified by the mapping exercise. It may be outlined as follows.

Table 7.1 Maps and satellite data used in the study

Map/satellite particulars	Sensor	Scale/resolution	Scene/map details	Year of survey/date of pass	Source
Police station map	—	1 inch ≡ 1 mile	Police station map of sagar	1922–23	Land Records Department, Government of West Bengal
IRS-1D	Pan	5 m	Path 109, Row 57	22 Jan 2001	National Remote Sensing Centre, India
	LISS-3	25 m	Path 109, Row 57, Quadrate A		
Sentinel 2A	MSI	10 m	Tile ID: T45QXE	04 Jan 2021	European Space Agency

7.2.1 RS-GIS Mapping Exercise

The details of maps and satellite images used in this study are shown in Table 7.1. These were processed to get an idea on the extent of erosion and accretion as the high tide line shifts in bordering the coastal *mouzas* (villages) of Sagar (Fig. 7.1). All the satellite data utilised were ortho-corrected on WGS-84 and UTM Zone 45.

Two Standard False Colour Composites (FCCs) were prepared from the 2001 and 2021 satellite data, which were interpreted visually for digitisation of the high tide lines. A Police Station Map of Sagar Block, depicting *mouza* (revenue village) boundaries was georeferenced and overlaid on the satellite FCCs to extract boundaries of the 42 *mouzas* located in the island. Global Navigation Satellite System positions of the sampled households were collected during field survey using a Garmin eTrex-20 receiver. These were exported as a point vector layer to the RS-GIS database and plotted showing the three main types of samples: displaced, resettled, and unaffected.

The Digital Shoreline Analysis System v.5.0, developed by the United States Geological Survey, was employed to estimate net shoreline movement (NSM) and end point rate (EPR). The NSM and EPR were found by computing the distance between the 2001 and 2021 shorelines for some 763 transects, positioned orthogonal to the two digitised coastlines (length: c. 75 km) with an interval of 20 years, dividing that distance with the time-lapse (20 years), and generating the predicted coastline for 2031. For estimating the location of the 2031 coastline, Long and Plant's (2012) Extended Kalman Filter was used.

7.2.2 Households Survey

The mapping task identified the major erosional *mouzas* (revenue villages) of the Sagar Island where about 90 households who have lost land to coastal erosion and

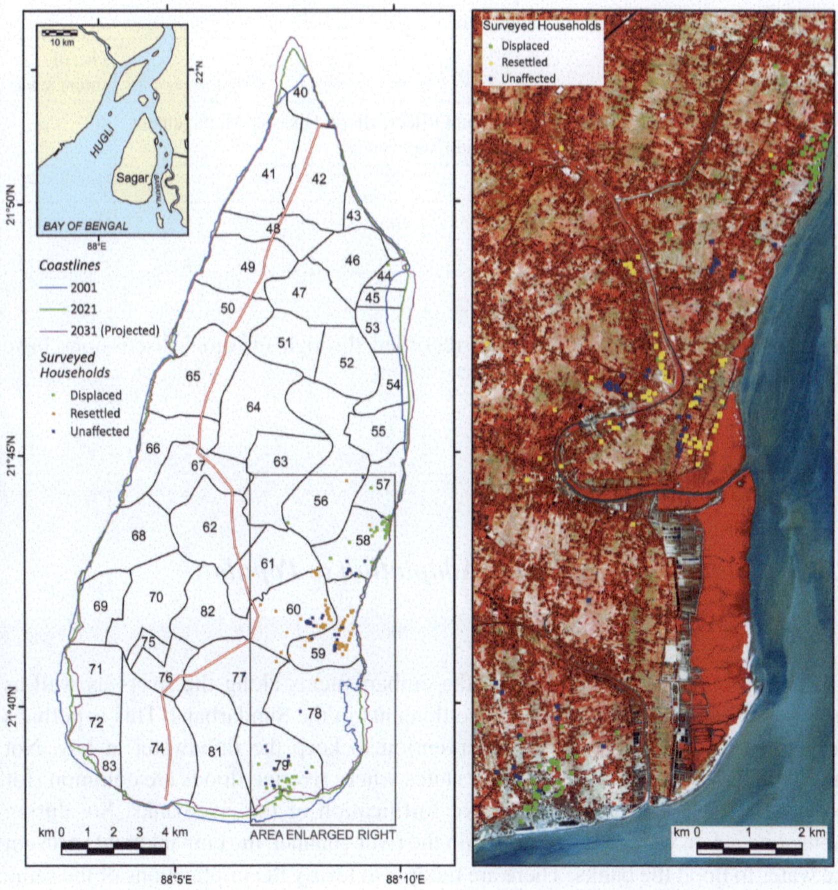

Fig. 7.1 Mapping physical vulnerability of coastal communities in Sagar Island. Numbers indicate *mouza* Ids. The enlarged map on the right is a Standard False Colour Composite of 2021. *Source* Fieldwork, 2021; IRS L3 + Pan merged data of 22 Jan 2001; Sentinel MSI 2A data of 04 Jan 2021; and Police Station map from Land Records Dept., Govt. of West Bengal, 1922–23

have not received any state support are designated as displaced households. Another 90 households were selected randomly from the Bankimnagar resettlement colony (designated as resettled). As control samples, 60 households were selected from the safer locations of the same mouza from where the 'displaced' and 'resettled' households were selected. Besides this, 13 in-depth semi-structured interviews were conducted the details are shown in Table 7.2.

The officials within the local governance who were chiefly responsible for disaster management in the study region were interviewed. The views of the *Panchayat Pradhan* (Chief of Village Council) and embankment sub-supervisor were especially important as their views were representative of the local people as well. From within the resettlement colony and other habitations surveyed, the hamlet leaders were

Table 7.2 Details of the interviews conducted

Nature of respondent	No. of interviews
Administrative officials (Block Development Officer, Block Disaster Management Officer, *Panchayat Pradhan*, embankment sub-supervisor)	6
Members of resettlement colony	3
Members of displaced households	2
Control sample households	2

included for interviews. Information recorded through informal discussions have also been incorporated in this study.

7.3 Results and Discussion

7.3.1 Embankment as (Mal)Adaptation to Populate the Sundarbans

Like many other flood prone areas, the embankments along the rivers as well as the seaface comprises the lifeline of settlements in the Sundarbans. This is perhaps the earliest form of engineering intervention to keep the floodwater at bay. Not only within West Bengal, but in other states where riverine floods are common (for example Bihar), there exists extensive fortification of the riverbanks. So, during monsoons, when water level rises within the river channel, the embankments prevent the water to flood the banks. There are records to testify the implications of the same are not too successful. This will emerge from the following discussion.

About half of the 104 islands in the Indian Sundarbans are presently inhabited (Danda et al. 2011; Bandyopadhyay 2021). Reclamation of these mangrove tidal forests for human habitation and agriculture was initiated during the colonial times with the sole objective of generating revenue out of the 'wasteland'. So, construction of some 3500 km of embankments along the tidewater interfaces and large-scale forest clearings were taken up in the immature deltaic islands. To the colonial government, there was little ecological value for the endemic mangrove forests which are not only important for carbon sequestration but are very vital 'first line of defence' against tropical cyclones. It became necessary that the regular ebb and flow of tidal water be restricted because now the erstwhile forest land are to be inhabited.

Since 1811, construction of mud embankments with wood pilling of about 5 m high were initiated around the Sagar. During high tides, water climbs up along the embankments to a level higher than in-island inhabited lands. It is obvious that a case of embankment breach will prove disastrous to the settled areas. Initially these were maintained by the *Zamindars* (landlords) and in post independent India the task was entrusted with the State Irrigation and Waterways Department in 1960

(Irrigation & Waterways Department, Government of West Bengal). Presently, the inhabited islands continue agriculture and aquaculture but at the mercy of the protection extended by these embankments. The maladaptation issues may be outlined as follows:

Firstly, the reclamation of immature deltaic land and their fortification prevents the regular tidal spillage. This stops the process of vertical accretion that is essential for raising the deltaic islands above the high tide line (Bandyopadhyay 2021). The reclaimed islands therefore continue to remain lower than the high tide level unlike the adjacent forest lands where high tide water inundates the island twice daily thereby depositing silt to add elevation to the islands. So, due to the embankments, the inhabited parts of the Sundarbans continue to remain geomorphologically immature and susceptible to tidal submergence (Fig. 7.2). With climate change induced sea level rise, the threats of submergence are even more severe.

Secondly, although the Sundarbans are a part of the abandoned component of the Ganga–Brahmaputra-Meghna Delta and receive only a small part of its sediment input, yet it sequesters about 10–15% of the sediments flux of the rivers that reach it besides the tide-borne sediments from the offshore (Bandyopadhyay et al. 2023). As the embankments debar tidal spillage and the subsequent spread of the sediment load on the deltaic islands, silt deposition occurs on the channel bed and water holding capacity of channels reduces systematically (Bandyopadhyay 2021). This happens because, in this region the rate of tidal rise is faster than tidal fall and this asymmetric pattern of the bidirectional tidal cycle turns the estuary into a sediment sink. This phenomenon makes it inevitable that the height of the embankments be raised regularly, or the channel beds be dredged to manage the channel capacity. The other aspect of the diminishing channel capacity and increasing height of high

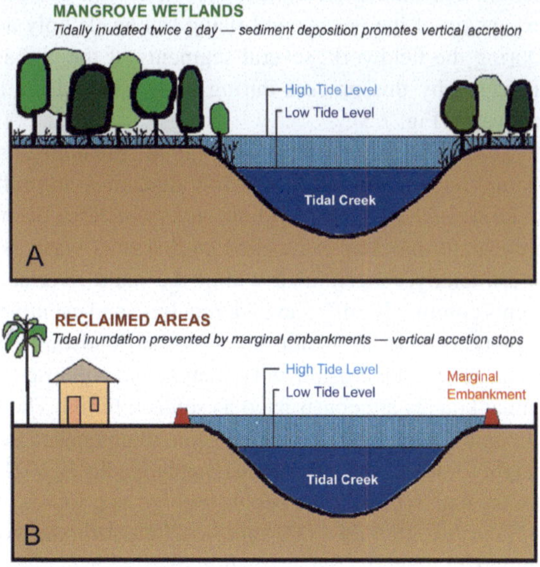

Fig. 7.2 Schematic diagram illustrating how placement of island-margin embankments in the reclaimed islands of the Sundarbans like Sagar effectively prevents sediment-laden tidewater from entering island interiors (**B**). In the non-reclaimed stretches (**A**), free-flowing tides deposit sediments and raise land level and can naturally negate the effects of rising sea levels. Based on Bandyopadhyay (2000)

Plate 1 Saline water stranded behind embankment

Plate 2 Partially breached ring embankment

Plate 3 Remnants of the wooden reinforcing structure after washing away of mud embankment

Plate 4 Process of embankment breach/overtopping during tidal surge

Fig. 7.3 Damaged embankments and areas inundated by saltwater

tide water is that the expanding water within the channel creates pressure along the walls of the mud embankments. Overtopping and/or breach of the embankments is a common phenomenon in the monsoon months and during the cyclones. The thermal expansion of the oceans and rising sea levels only add to the existing complexity. During the fieldwork, several segments of the embankments were observed to be damaged by this process during high tide and the area was found inundated by saltwater (Fig. 7.3).

Thirdly, the steep-sided (~45°) mud embankments create an additional paradox of interfering with wave and flow dynamics and reflect the waves to cause basal scouring of the dykes. The tidal creeks which run deep inland within the settled areas are usually blocked or fortified so that tidal water is kept at bay and tidal current is artificially concentrated within the main riverine channels only. The embankments commonly suffer an undercut base and eventually collapse. Plate 2 of Fig. 7.3 depicts how coastal erosion has proceeded landward destroying the marginal embankment. As coastal erosion and embankment collapse continue unabated, parallel ring embankments are constructed as setback limits. The embankment emerges as sites of contestation and require consistent maintenance in the form of repair and raising height for long term adaptation (Bandyopadhyay 2021). Thus, the embankments that are created to protect the people emerge as a source of their vulnerability.

Fourthly, after the 2009 super-cyclone *Aila*, the Government of West Bengal has embarked upon the reconstruction of marginal embankments for 177 km stretch of washed-out/damaged and 601 km of severely damaged parts (Irrigation & Waterways Department, Government of West Bengal) using concrete and a low-slope design. This scheme was techno-economically approved and proceeded through subcontracting to private builders through call of tender. It is common to find more than one contractor working along the same stretch and joining their own components with the subsequent contractor's segment. These points of juncture remain the points of weakness. The interviews reported that corruption and use of low-grade construction material makes the embankments not as strong as they were intended to be. The respondents reported that disputes remain regarding land acquisition for the redesigned embankments and that the scheme has proceeded leaving aside the disputed stretches of land (which are nonetheless essential to complete the structures) in interesting ways that defeats the purpose. Thus, the marginal embankments are constructed using highly sophisticated engineering technologies but left gaps in between as some stretches of the marginal land has property rights issues and cannot be acquired. These gaps work as doorways for tidal surges and nullify the huge investment made for upgrading the embankments.

Fifthly, the storm surge overtopping of embankments especially during the severe tropical cyclones have become common incidents. Not only saline water ingress destroys crops and freshwater ponds, but also the water is held back behind the embankments for prolonged periods in absence of drainage mechanism (Plate 1 of Fig. 7.3).

In short, the embankment issue is ridden with interesting contradictions that are both geomorphological as well as socio-economic. The climate change induced sea-level rise and increase in the incidence of severe tropical cyclones have exacerbated the pre-existing vulnerabilities. The embankments as a strategy to populate the Sundarbans have assumed complex shades because of both the ecology and climate induced vulnerabilities.

7.3.2 Population Relocations as (Mal)Adaptation to Shrinking Islands

Taking note of the rapidly eroding Lohachara and Ghoramara islets within the Sagar CD Block, it was in 1964 about 25 households were officially relocated to Sagar Island at Phuldubi and South Haradhanpur (Chakma and Bandyopadhyay 2012; Montreux et al. 2018). Thereafter, large-scale resettlement of displaced population of these islands were undertaken at Bankimnagar (1972), Gangasagar (1981) and Jibantala-Kamalpur (1983) when the Communist Party ascended to power in West Bengal. The Beguyakhali resettlement colony is the most recent one, launched in 1994–95. The District Councils and Village Panchayats were implementation agencies. These projects were undertaken in the freshly reclaimed intertidal areas in Sagar and part of this was funded by the *Indira Awas Yojana* scheme (Bandyopadhyay 1997). The land allocation gradually declined from about 0.8 hectare including allocation for homestead in the Bankimnagar Colony (1972) to 0.2 hectares at the Jibantala-Kamalpur colony (1983). This land allocation was undertaken legally through issue of *patta*—an instrument developed during the Land Reforms to extend land to tillers of land. The resettled population are legal residents of the region having formal land titles distributed as part of the resettlement scheme. It also entailed monetary allocations for construction of cyclone proof homesteads.

The interviews at the Bankimnagar resettlement colonies during the field survey in February 2021 revealed that the resettled population seems to be grateful for whatever government support they have received appreciating the fact that there is scarcity of government land available for distribution (also see Montreux et al. 2018). The Government had taken complete responsibility of relocating the people who were willing to accept support from the eroding Ghoramara Islet to Sagar resettlement colonies. Recently, the panchayat areas of Gangasagar and Dhablat have been brought under the jurisdiction of the Bakkhali Development Authority (BDA). So, execution of *patta* for any land requires approval of the BDA. Hence corruption related to the access to land for resettlement is claimed to have been controlled to some extent (Interview with Deputy to Panchayat Chief, i.e. Upa-Pradhan, Dhablat Panchayat dated 10th February 2021).

The prime issue of concern centres on the geomorphology of the Sagar Island itself. The displaced population of Lohachara and Ghoramara were resettled in Sagar Island mainly to keep them within the familiar ecology and administratively within the same CD Block. But the prevalence of severe erosion in several parts of the latter was overlooked. Figure 7.1 depicts the location of the resettled households (yellow points in Fig. 7.1). They lie all along the ring embankments mostly. It is already discussed in the previous section how nearness to embankment can be a potential threat from storm surge and embankment breach. There exist several studies that point out the geomorphic fragility of the Sagar Island itself not only due to the unavailability and reworking of sediments under tidal and wave action (Bandyopadhyay 1997; Bandyopadhyay et al., 2023) but also due to the climate change induced sea level rise (Hazra et al. 2002; Danda et al. 2019) and auto-compaction related subsidence (Bandyopadhyay 1997).

The Bankimnagar Colony and the Gangasagar Colony near Beguyakhali are two zones that are consistently eroding severely such that the colony presently stands within less than a kilometre from the coastline. The fieldwork sought to map the location of the households and Fig. 7.1 clearly brings out the fragile location of the resettled population. These communities have already been uprooted due to severe coastal erosion arising out of the interplay of coastal processes and sea-level rise, and now they have been resettled in a location which is not only susceptible to coastal erosion but also subjected to degeneration of the ecological processes in the resettlement colony areas. Further, Bandyopadhyay (1997) points out that reclamation of the intertidal zones in the coastal environment is open to erosional processes and that anthropogenic interference undeniably would lead to deterioration of the intra-island creeks. Experts doing research in these issues clearly proclaim planned relocation of population away from Sagar Island to the more stable northern parts and to allow the coastal process to reclaim land back into the coastal dynamics for it to stabilize

(Danda et al. 2011; Danda et al. 2019). The present district disaster management plan however does not recognize the imperceptible yet pervasive threat to land livelihoods in Sagar and has no mention or any recognition of the need to take any action towards planned relocation of the communities. This apparent invisibility of this major threat in the disaster management documents perhaps deepens the crisis and makes redressal all the more challenging.

To understand the broad implications of the resettlement exercise, the main source of income of the respondents was recorded. Figure 7.4 depicts the principal source of income of the three communities. Firstly, both agricultural activities and non-agricultural enterprise-based activities are relatively more important mode of livelihood for the unaffected households in comparison to the displaced and resettled population. Secondly, wage employment, which is the most precarious mode of livelihood, is the dominant mode of income for most of the displaced and resettled households. A comparison of their monthly per capita income (MPCE) as a proxy for income revealed that the MPCE of the displaced (₹ 1609) and the resettled communities (₹ 1614) are not significantly different but that of the unaffected households (₹ 2025) is significantly higher (Table 7.3). Economically there exits little difference between the displaced and the resettled communities and some members of the latter group reported their experience of deteriorating condition. Their loss of social capital in addition to their land and other assets have not been replenished adequately to help them to bounce back through resettlement. It highlights that the unaffected communities are relatively better-off compared to the displaced and resettled communities. The more critical finding relates to nearly equal vulnerability of the displaced communities and those resettled households who are recipients of State support. Informal interviews have reported some deterioration in the social and economic condition of the resettled people in course of their relocation into the Sagar resettlement colonies.

7.4 Conclusion

The Indian Sundarbans in the Bengal Delta is an interesting locale where developmental deficit coexists with environmental stress and climate (mal-)adaptation instances. The most important technique for adaptation, i.e., the embankments, emerged in the region two centuries ahead of any discourse on climate adaptation or disaster studies. However, the lived experiences of the respondents have clearly underlined that this method is highly flawed and that it interacts with the natural processes resulting into negative outcomes. It makes the settlements vulnerable even to a normal high tide not to speak of the storm surges. A clear understanding of the geomorphology of the region emerges as a pre-condition for the sustainability of the socio-ecological systems. At some point therefore it may sound deterministic in the way that environment tends to exert the long-term control over human society. It is presented elsewhere (see Mallik et al. 2023) how institutions have evolved and have been eventually modified in response to environmental crisis. The experience of the resettlement scheme may be located within the larger discourse on climate

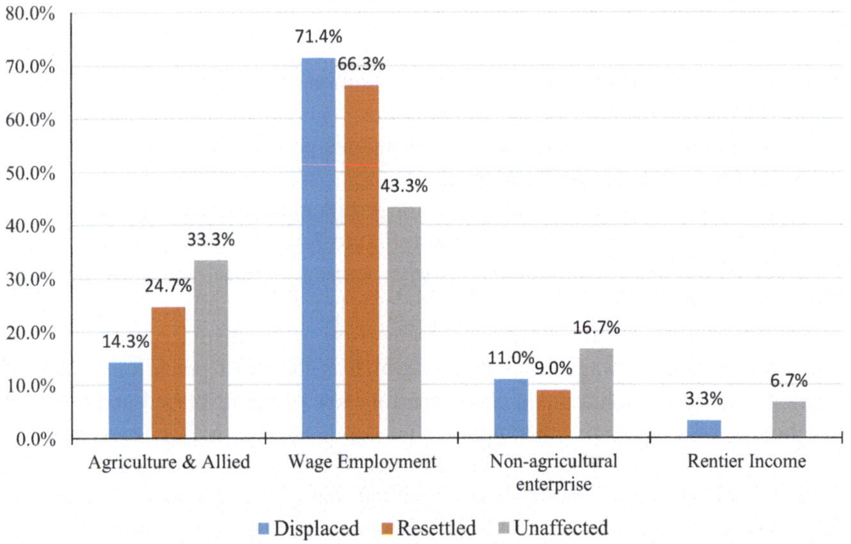

Fig. 7.4 Principal sources of income of the sampled households (n=232). *Source* Fieldwork, 2021

Table 7.3 Comparison of monthly per capita expenditure (MPCE in rupees)

	Characteristic of households	N	Mean
Category of household	Displaced	87	1609
	Resettled	88	1614
	Unaffected	57	2025
	Total	232	1713

Note Mean MPCE of unaffected households is significantly higher than the others at 5% level
Source Fieldwork, 2021

induced planned population relocation. While existing literature categorises such population mobilities as voluntary, involuntary, and trapped populations, an analysis of their correlates brings out the multi-causal nature of mobility decisions and their complexities.

Although the recent IPCC reports as well the COP sessions recommend that the national governments must consider planned population movement away from coastal areas and other spaces where environmental crisis is imminent, it is ridden with questions of re-establishing of social and cultural artefacts besides questions of property and livelihoods (McAdam and Ferris 2015). Very few countries have, therefore, been proactive to embrace planned relocation as a measure for climate adaptation as it raises more questions than it resolves. In consonance with most of the studies on resettlement that report general deterioration of the migrants' lives, this study reiterates similar experience. Besides issues that mark socio-economic sufferings, it brings out the concerns around increased population pressure in a region which itself is threatened by erosion and witnesses distress induced outmigration as a major livelihood strategy. The selection of the location of some of the resettlement colonies in the recent accretion zones emerge particularly concerning as it disrupts the land building process in the deltaic environment. Hence, both the strategies of peopling the Sagar Island clearly appear maladaptive. What is more concerning is the insensitivity of the policy makers regarding the long run implications in the context of rising sea level and increased frequency of coastal storm surges. While the proposition of 'planned population relocation' of Danda et al. (2019) apparently seems to be environment friendly, its long term social, political, economic, and not the least environmental outcomes must be carefully understood. In the present form, the study region is doubly vulnerable as a result of the long-standing adaptation measures which have proved to be maladaptive.

Acknowledgements This study is funded by the Indian Council for Social Science Research (ICSSR) under the Impactful Policy Research in Social Science (IMPRESS) scheme vide Project Sanction No: IMPRESS/P970/283/2018-19/ICSSR. The authors are thankful to Prof. Sumana Bandopadhyay, University of Calcutta, for her support. Pritha Boral, Piyali Dev and Daibosree Adak are acknowledged for their help in the fieldwork and data aggregation tasks. Pritam Kumar Santra helped to prepare the maps.

References

Bandyopadhyay S (1997) Natural environmental hazards and their management: a case study of sagar island. Singap J Trop Geogr 18(1):20–45

Bandyopadhyay S (2000) Coastal changes in the perspective of long term evolution of an estuary: Hugli, West Bengal, India. In: Rajamanickam V, Tooley MJ (eds) Quaternary Sea Level Variation. Shoreline Displacement and Coastal Environment, New Academic Publishers, Delhi, pp 103–115

Bandyopadhyay S (2021) The reclaimed Sundarban in a warming world: A short geomorphic appraisal. In: Chakma N, Jana NC, Siddique G, Guchhait SK, Samanta G, Mistri B (eds) Resonance of Environment, Society and Space: Global Perception through Regional Perspective, Golden Jubilee Commemorative, vol of. Department of Geography. The University of Burdwan, La Strada Publishers, Kolkata, pp 31–45

Bandyopadhyay S, Kar NS, Dasgupta S, Mukherjee D, Das A (2023) Island area changes in the Sundarban region of the abandoned western Ganga–Brahmaputra–Meghna Delta India and Bangladesh. Geomorphology 422:108482. https://doi.org/10.1016/j.geomorph.2022.108482

Black R, Arnell NW, Adger WN, Thomas D, Geddes A (2013) Migration, immobility and displacement outcomes following extreme events. Environ. Sci. & Policy 27s:s32–s43

Chakma N, Bandyopadhyay S (2012) Swimming against the tide: survival in the transient islands of the Hugli Estuary. West Bengal, West Bengal Geo-spatial Issues, Department of Geography, University of Burdwan, West Bengal, pp 1–19

Christian-Smith J, Morgan CL, Peter JG (2015) Maladaptation to drought: a case report from California, USA. Sustain Sci 10(3):491–501. https://doi.org/10.1007/s11625-014-0269-1

Danda AA, Sriskanthan G, Ghosh A, Bandyopadhyay J, Hazra S (2011) Indian Sundarbans delta: a vision. World Wide Fund for Nature-India, New Delhi

Danda AA, Ghosh N, Bandyopadhyay J, Hazra S (2019) Managed retreat: adaptation to climate change in the Sundarbans ecoregion in the Bengal Delta. J. Indian Ocean. Reg., 15(3): 317-335. https://doi.org/10.1080/19480881.2019.1652974

Dasgupta S, Wheeler D, Bandyopadhyay S, Ghosh S, Roy U (2022) Coastal dilemma: Climate change, public assistance and population displacement. World Dev 150:1–14. https://doi.org/10.1016/j.worlddev.2021.105707

Diana R, Magnan AK, Singh C, Lukas-Sithole M, Ben Orlove E, Schipper LF, Coughlan E, de Perez, (2023) Navigating the continuum between adaptation and maladaptation. Nat Clim Chang 13:907–918. https://doi.org/10.1038/s41558-023-01774-6

Edwards, J (2016) A story of relocation and rising sea levels. Vunidogoloa Village, Vanua Levu, Fiji. New World Outlook: The Mission Magazine of the United Methodist Church.

Entzinger H, Scholten P (2016) Relocation as an adaptation strategy to environmental stress: lessons from the mekong river delta in Vietnam. IOM Migr, Environ, Clim Chang Policy Brief Ser 1(6). https://publications.iom.int/system/files/pdf/policy_brief_issue6_1.pdf

Ferris E, Weerasinghe S (2020) Promoting human security: relocation as a protection tool in a time of climate change. J. Migr. Hum. Secur. 8(2):134–149

Georgetown Climate Center (2019) GCC Hosts interdisciplinary convening exploring managed retreat as a strategy for community adaptation. https://www.georgetownclimate.org/articles/gcc-holds-2-day-workshop-onmanaged-retreat.html

IPCC (2022) Climate change 2022. Impacts, adaptation, and vulnerability. Contribution of Working Group II to the Sixth assessment report of the intergovernmental panel on climate change

Irrigation & Waterways Department, Government of West Bengal accessed at Irrigation & Waterways Department (wbiwd.gov.in) on 23/09/2023

Juliet C-S, Levy Morgan C, Gleick PH (2015) Maladaptation to drought: a case report from California, USA. Sustain. Sci. 10(3):491–501. https://doi.org/10.1007/s11625-014-0269-1

Kothari U (2014) Political discourses of climate change and migration: resettlement policies in the Maldives. Geogr J 180(2):130–140. https://doi.org/10.1111/geoj.12032

Long JW, Plant NG (2012) Extended Kalman Filter framework for forecasting shoreline evolution. Geophys Res Lett 39(13):1–6

Mallik C, Bandyopadhyay S, Bandopadhyay S (2023) Land scarcity and land access in a hazard-prone island: Sagar, Indian Sundarbans. Singap J Trop Geogr 44:255–276

McAdam J, Ferris E (2015) Planned relocations in the context of climate change: unpacking the legal and conceptual issues. Camb J Int Comp Law 4(1):137–166

Mortreux C, Safra R, de Campos WN, Adger TG, Das S, Adams H, Hazra S (2018) Political economy of planned relocation: a model of action and inaction in government responses. Glob Environ Chang 50:123–132

Piggott-McKellar AE, Nunn Patrick D, McNamara KE, Sekinini ST (2020) Dam(n) seawalls: a case of climate change maladaptation in Fiji. In: Leal Filho W (ed) Managing climate change adaptation in the pacific region. Springer Nature, Switzerland, pp 69–84. https://doi.org/10.1007/978-3-030-40552-6_4

Pritchard B, Thielemans R (2014) Rising Waters Don't Lift All Boats': a sustainable livelihood analysis of recursive cycles of vulnerability and maladaptation to flood risk in rural Bihar, India. Aust Geogr 45(3):325–339. https://doi.org/10.1080/00049182.2014.930001

Reckien D, Magnan AK, Singh C, Lukas-Sithole M, Orlove B, Schipper ELF, de Parez EC (2023) Navigating the continuum between adaptation and maladaptation. Nat Clim Change 13(9):907–918. https://doi.org/10.1038/s41558-023-01774-6

Sugata H, Ghosh T, DasGupta R, Sen G (2002) Sea Level and associated changes in the Sundarbans. Sci Cult 68(9–12):309–321

Chapter 8
"Maladapted" Public Transport Solutions: A Case of Amritsar in Punjab, India

Kanchan Gandhi and Raman Sharma

Abstract Amritsar is a historical, cultural, and spiritual center in Punjab. The old core of the city is characterized by narrow lanes that are accessible by Intermediate Public Transport (IPT) modes only. Hence, diesel-fueled auto rickshaws and battery-operated rickshaws along with private vehicles are the dominant modes of transport in the city. The government of Punjab launched a Bus Rapid Transportation (BRT) system in Amritsar in the year 2014 in an attempt to boost the mass transit system in the city. The BRT has however largely been termed as a "failure" by the people of the city since it is currently operational only on three routes and there is still heavy reliance on IPT to commute within and outside the city. The other scheme that was aimed at climate change mitigation by reducing carbon emission was the phasing out of diesel auto rickshaws and switching to electric auto rickshaws seems like a "partially maladapted solution." The scheme includes a subsidy component to facilitate the shift to "cleaner energy." However, power in Punjab is largely generated in coal-based thermal power plants. The share of renewable energy is relatively low in the state. Therefore, shifting to e-auto rickshaws would simply mean shifting from one fossil-fuel-based energy to another. This chapter will evaluate the two transport schemes—the BRT and the e-auto rickshaw scheme to argue that these are examples of maladapted solutions to climate change mitigation since they are currently not integrated with each other; and give recommendations to make them environmentally sustainable and better adapted to reduce carbon emission.

Keywords BRTS · E-auto rickshaws · Maladapted solution · Integrated public transport system · IPT · Climate change mitigation

---

K. Gandhi (✉)
Independent Researcher and Visiting Faculty Member, School of Planning and Architecture, Delhi, India
e-mail: kanchan.gandhi1977@gmail.com

R. Sharma
Advocacy Officer, Federation of Indian Animal Protection Organisations, Bhathinda, India

A. Sarkar et al. (eds.), *Risk, Uncertainty and Maladaptation to Climate Change*, Disaster Risk Reduction, https://doi.org/10.1007/978-981-99-9474-8_8

8.1 Introduction

In this chapter, we attempt to describe and analyze two transportation interventions in the city of Amritsar in Punjab—the Bus Rapid Transportation systems (BRTS) project and the "Rejuvenation of Auto Rickshaws in Amritsar through Holistic Interventions (RAAHI)" e-autorickshaw scheme. The purpose of analyzing these schemes is that both of these are purportedly aiming the reduction of emissions by enabling people to switch to public transportation and towards the usage of cleaner fuels but in their present form they are both partially maladapted solutions. The analysis below will demonstrate how these schemes are only partially effective in the reduction of emissions and propose how they can be made more sustainable in the future.

The BRTS was a politically motivated project that was pushed by the government of Punjab in the year 2014 under the Jawaharlal Nehru National Urban Renewal Mission (JNNURM) introduced by the central government in the year 2005. During interviews conducted by the authors, the professionals in Amritsar—planners and architects alluded that there was no public consultation for this project. This project has had only a limited impact on the city's mobility pattern and reduction of pollution because it currently (in 2023) runs only on three routes on a stretch of 31 km and the buses are currently diesel-fueled. Due to its limited scope, it did not motivate people to make the shift from private vehicles and Intermediate Public Transport (IPT) options to bus transportation. We argue that because the project lacked public participation, ran on diesel-fueled buses and did not connect all parts of the city, it is currently maladaptive to solve the problem of high carbon emission. Moreover, it has not led to the promotion of the use of mass public transportation services or the reduction of reliance on private vehicles. In the last section, we give recommendations to make this project more popular and climate-friendly so that people will shift toward public transportation eventually.

The second project that we discuss in this paper is the RAAHI e-autorickshaw scheme that is funded by the French development agency and partly by the Smart Cities Mission and targets the reduction in the fleet of diesel-fueled auto rickshaws by distributing e-auto rickshaws to drivers in the city. Since e-auto rickshaws are more expensive than the diesel ones, there is a subsidy component offered to the drivers to make a switch to the "cleaner" energy option. According to the scheme implementing team, this was a well-intended and conceptualized scheme which faced initial launching problems such as hesitance among auto-drivers to switch to e-vehicles due to high cost of the e-auto rickshaw and lack of charging infrastructure on major tourism routes. The implementation team pushed for the increase in the subsidy component for the e-auto purchase, facilitated the process of easier access to bank loans, and developed charging infrastructure across the city to facilitate the switch to e-auto rickshaws. The e-auto rickshaw scheme runs on electricity generated largely from fossil fuels instead of renewable sources of energy, making it a maladapted solution to climate change mitigation.

8.2 BRTS—Challenges in Indian Cities

Indian cities—big, medium, or small in size struggle with transportation issues as buses, cars, cycles, trucks, and pedestrians compete for space on the roads. The population pressures on cities are often reflected in the chaos on the road. Additionally, India is home to some of the cities with the highest air pollution levels in the world. Bus transport systems were emphasized during the urban renewal program, JNNURM launched by the Government of India in the year 2005. Under this program, there was an emphasis on strengthening public transport systems, particularly city buses. A new fleet of buses was procured and added to most cities selected under this scheme. According to a press release by the Government of India dated 13th February, 2014 (GoI, 2014), BRTS projects were sanctioned across 11 cities in the country including Pune, Indore, Bhopal, Jaipur, Ahmedabad, Surat, Rajkot, Vijayawada, Vishakhapatnam, Amritsar, and Kolkata. Nine years after these projects were sanctioned, built, and operationalized most of them have been termed as "massive failures." For example, in Bhopal the residents are demanding the dismantling of the BRT infrastructure as it has been a cause of serious accidents in the city. In other cities as well, the BRT has caused huge loss of public money as city residents have not switched to it in desired numbers. Delhi is another example where the residents rejected the BRT system, and it was dismantled causing huge financial losses.

In the case of Indian cities, the BRT seems to be largely a maladapted solution since the roads already experience high volumes of traffic. The BRT dedicated lanes lead to the narrowing of the existing roads for other vehicles, which in turn try to encroach the BRT lanes leading to accidents and chaos on the streets. Therefore, the residents of most cities where the BRT was approved have rejected the system. Moreover, most cities that implemented the BRT bought a fleet of diesel run buses which did not contribute to emission reduction. A report by the Observer Research Foundation concluded that due to the various challenges faced in its implementation the BRTS seems to have been overshadowed by its more glamorous kin—the metro (Jha 2020).

One of the reasons for maladapted transport solutions in cities of the Global South is due to high political interference in the selection and implementation of projects. For example, transport expert Tiwari (2020) argues that the BRT is a much better suited solution to the traffic and transportation issues in India. However, the central government's focus on the metro rail systems has caused even the smaller cities to aspire for this system. The metro rail system costs approximately Rs 300 crores/kilometer (USD 37 Million) as compared to Rs 12 crores/kilometer in the case of Amritsar BRT (USD 1.4 million) for the BRT. A capital-intensive project like the metro rail is not feasible for medium-sized cities like Amritsar that have a population of around two million.

The project managers and bureaucrats seldom consult people during the formulation of projects. They are under political pressure to implement their favorite projects of the politicians. There are several examples of maladapted projects, greenfield cities that ended up as urban disasters in India (Sampat 2016; Kolsto 2017). Vote-bank politics and lobbying by private operators in public transportation hinder the implementation of public transport projects like the BRT.

8.3 BRTS—Success Stories

One of the most successful cases of the BRTS in ameliorating traffic, transport, and pollution problems is Bogota in Colombia. An analysis by Hidalgo et al. (2013) illuminates that the Bogota BRT made a big positive impact in the city's mobility pattern. It helped lower income people to travel cheaply. The network of the BRT connected to most parts of the city and was well integrated with feeder services and IPTs. It helped in reducing the air pollution, crimes, accidents and improving investments and land values in the city.

In India, Ahmedabad is considered as a success story of the BRTS. The ring and radial pattern of the BRT links the core of the city with its periphery in all directions (Fig. 8.2). It connects to all the activity centers and thus is highly used by residents of the city (Jaiswal et al. 2012). A report by Rana (2022) highlights that in Ahmedabad 20–22% of the commuters have moved from using their motorcycles to the bus. With an average trip length on the bus of 7 km, this translates into a saving of almost 200,000 vehicle kilometers per day (5,000,000 per month). The report concludes that the BRT led to the overall rejuvenation of the city since it connected most parts of the city and covered maximum destinations. Parking facilities are provided near most BRT stations in Ahmedabad that allow people to park their private vehicles thus taking care of the last mile connectivity issue. Ahmedabad is one of the success stories of the BRTS in India. It was the first BRT to get a fleet of electric buses in the year 2021, thereby making it a greener transit option.

Like Ahmedabad, Amritsar too has a similar radial pattern of roads emanating from the core toward different directions in the periphery (Fig. 8.1).

The BRT system needs to be expanded to these different radial routes and integrated with the IPTs in the core of the city which has narrow roads not suitable for a bus system. The outer core of the city needs a fleet of minibuses to integrate with the bigger buses in the periphery of the city. In the case of Amritsar, however, only the first phase of the BRT has been implemented. The feeder routes have not yet been developed and last mile connectivity is not possible with the incomplete service. Hence, the Amritsar metro service elicited mixed user satisfaction.

The BRTS in Istanbul, Turkey too has elicited a mixed response from the users and experts. While the ridership of the BRT is quite high in Istanbul, some experts believe that the BRT should have been developed with rail technology (Babalik-Sutcliffe and Cengiz 2015). The authors in this chapter argue that success of BRTs can be replicated in other cities only with careful planning.

Fig. 8.1 Road map of Amritsar. *Source* Road map of Amritsar, Maps of India, retrieved from https://www.mapsofindia.com/maps/punjab/roads/amritsar.htm

8.4 Methodology

The fieldwork for this chapter was conducted from April 2022 to July 2023. Interviews were done with key stakeholders in the transport administration of Amritsar city. Interviews were conducted with the key government officials including the project manager of the BRTS in Amritsar and Chandigarh, project consultants of the two projects, auto drivers and professionals in the city.

An online perception survey of residents about the BRTS was conducted in June 2023. The results of this survey were used to identify the main issues in BRT usage stated by the respondents. The respondents of this survey included professionals, University students, teachers and homemakers. The sample size of the online survey was 51 persons mostly from middle class. Apart from the primary surveys, secondary data was collected from the Detailed Project Reports and feasibility reports of the two schemes: from the Comprehensive Mobility Plan and Master Plan of Amritsar, 2031. Analysis of newspaper articles was done to understand the politicization of the two projects.

Fig. 8.2 Ahmedabad BRT network, retrieved from https://upload.wikimedia.org/wikipedia/commons/4/47/Ahmedabad_BRTS_Network_Map.png

8.5 Amritsar—A Mobility Profile

The Bus Rapid Transit System (BRTS) or Amritsar Metro Bus was partially launched in 2016 and in 2019 it was launched with its full capacity. The project cost approximately 54.5 million rupees (650 thousand dollars). The BRTS project was commenced to reduce the traffic congestion in the city and to provide a better public transport system for the commuters. As the city's economy is dependent on tourism, the existing 31 km long BRTS system is inadequate as does not connect any tourist places in the city. Auto rickshaws are the only option for tourists to commute within the city.

The Comprehensive Mobility Plan (CMP) for Amritsar, 2012 emphasized the preparation of the Transportation model with a predicted share of 40% by Public Mass Transport System (PMTS) and recommended development of Integrated Multi Modal Transport Plan for 20 years with 4 phase development. The CMP identifies the lack of public transportation and the heavy dependence on private vehicles and auto rickshaws as the main mobility problem in the city. It says that 99% of the vehicles in the city are small occupancy ones which cause congestion and pollution in the city. The percentage of expenditure on transport in Amritsar is 15% of the household income as compared to 10% in the cities where public transportation options are available (CMP, Amritsar 2012).

The feasibility report for the RAAHI project (2020) highlights the importance of IPT in Amritsar:

> IPT in Amritsar is primarily based on a concept of shared mobility and therefore plays the crucial role of providing a low-cost option for commuter movement across the city. The IPT sector has mushroomed over the past 15 years in Amritsar and is currently very disorganized leading to congestion at popular locations while lack of service at others, along with high levels of air and noise pollution, and unsustainable competition among drivers leading to lower incomes. (RAAHI project feasibility Report 2020)

The report further identifies mobility as the key problem in the city due to the rise in the number of illegal and non-registered rickshaws that have resulted in extreme traffic conditions in core areas of the city. This has further led to poor uptake of public transportation resulting in increased private vehicles. Hence, there is an urgent need for formalization of the IPT sector in Amritsar city. The administration in Amritsar from the year 2022 exacerbated its efforts to create awareness to convince and enable the diesel auto-rickshaw drivers to switch over to e-auto rickshaws. The tagline for the RAAHI project is "Behtar Kamai, saaf vaatavaran" (better earning, cleaner environment).

Presently the auto rickshaws in the city dominate the public transportation in the city and compete with the buses and the BRT system as they operate on all the major tourism circuits.

8.6 BRT in Amritsar

The first phase of the BRT operates on the following routes as shown in Fig. 8.3 with entry Gate (Golden Gate) to India Gate/Back, Verka to India Gate/Back, Verka to Entry Gate/Back. The entry gate is the entry point to Amritsar city on the Jalandhar-Amritsar Highway and the exit to the city is toward India Gate that comes before Wagah border. This limited outreach of the bus network hinders people in the city from using it. According to the survey conducted by the authors the residents want more BRT routes to be operational that will connect the city to its periphery and particularly to the airport.

The BRT was a project initiated under the JNNURM scheme during the Shiromani Akali Dal (SAD) and Bharatiya Janata Party (BJP) coalition government in Punjab in the year 2014. However, the BRT and the Heritage augmentation scheme (HRIDAY) projects launched by the SAD government failed to capture the vote banks that they had hoped, and they lost the state government election held in the year 2017 in which the Congress party emerged victorious.

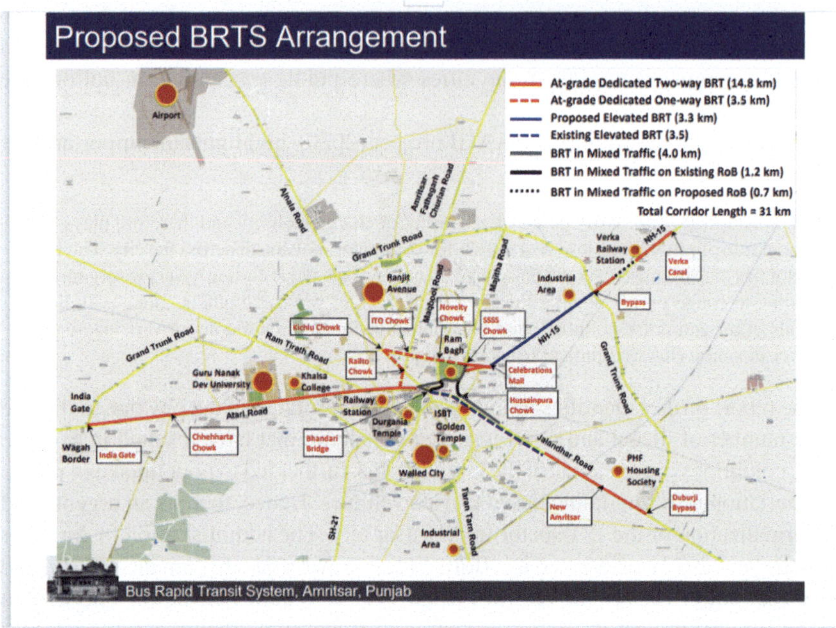

Fig. 8.3 Route map of the first phase of BRT. *Source* Taken from the BRT detailed project report. Reproduced with permission from Punjab Bus Metro Society (PBMS)

Initially the BRTS was operated by Punjab Infrastructure Development Board (PIDB), but later on it was handed over to Punjab Municipal Infrastructure Development Company (PMIDC) and now the Amritsar Municipal Corporation manages it. In September 2019, the Punjab government levied 10 paise per liter additional tax on high-speed diesel and petrol within the urban areas of the state for Punjab urban transport fund. This fund has been established for urban transport development, viability gap funding and infrastructure for urban transport projects recommended or approved under the various schemes of the state government and the Union Government.

A professional in the city said that "BRT is the white elephant in the room, a total waste of resources, it makes huge losses. It is true that the operation of BRT in Amritsar ran into several issues including the bankruptcy of the private operator in the month of June 2023 who was in charge of running the service in the city. The private operator said that they were not paid by the government on time for operating the BRT in Amritsar and hence were facing problems in paying the staff including the bus drivers and the ticket vendors. The users of the BRT also said that the network of the BRT needs to be expanded to cover all parts of the city to make it more usable. The absence of last mile connectivity deters many people from taking the bus. The working male especially prefers to take auto rickshaws or use their personal bikes to get to work. The bus is popular among school and college students and women who have more time at their disposal to get to places.

It is notable that the narrative around the BRTS was remarkably different between the bureaucrats managing it and the residents of the city. During the interviews, the project manager of the BRTS in the city iterated that the BRT was a success till now since around 45,000 people use it every day. However, the residents of the city said that it was not very useful due to crowded buses, limited route, lack of last mile connectivity, and the competition from auto rickshaws.

A new BRTS route was started between Gheo Mandi (near Golden Temple) and Amritsar Airport in 2021. This new route had increased the ridership, but private mini-bus operators opposed the move. The Tribune (2021) reported that after some days the route was stopped due to political interventions and despite the demand of local citizens and civil society, the government didn't make any effort to resume it. Authorities had also suggested extending the route of BRTS for Ram Tirath Road and also providing service to Attari-Wagah Indo Pak border. However, no decision had been taken as of July 2023. Currently the BRT operates only within the city limits. In an interview, the project manager of the BRTS in Amritsar (interviewed in June 2023) alluded that the per day ridership of the BRT in 2023 was around 40,000 which had increased from 2021 (30,000) due to the awareness campaigns conducted in universities, public parks, and schools. School children were given free passes to ride the BRT. He also said that the city administration had sent a proposal to the state government for constructing phase 2 of the BRTS which will entail the development of the feeder bus services and the airport route. He said that they had run into problems with the private operator who was given the contract for operating the BRT in Amritsar.

Transport experts have argued that the BRT systems that are spread over the city are more successful than those running on limited routes. Presently only Phase 1 of the BRT has been implemented. The residents want more routes to be operational in the city.

The Tribune (March 2023) quoted Kunwar Vijay Pratap an Aam Aadmi Party MLA from North (Amritsar Urban) who called the BRTS a "flop show" alleging that the "SAD-BJP government had ruined 55 million rupees on this project. The BRTS lane in the middle of the road creates routine traffic congestion on the city roads. As a result, commuters face long traffic jams on a daily basis in the city." Despite the critiques, BRT has the potential to strengthen the public transportation system in Amritsar if some measures are taken to strengthen the system. By implementing phase 2 of the bus network, Amritsar can address the problem of congestion and air pollution that is rampant in the city.

In India, it is typical that the ruling party leaders criticize the projects selected and implemented by their predecessors. This has a negative impact on future investment in that infrastructure and sometimes causes its dismantling and collapse which is counterproductive to growth. The assertion of the AAP MLA that the BRT was a waste of resources is challenged by the arguments of transport experts. The emissions from the BRT are much less than the metro train system if the electricity required for the train is generated from fossil fuels. The life-cycle analysis of the transportation system is done by Tiwari (2020) from manufacturing, usage and disposal of buses and a comparison of the GHG emission per passenger kilometer is 14.1 for bus system and 24.2 for metro system.

In small and medium cities, the BRT is a better option than the trains. However, the anomaly in India is that many of the cities in India where the BRT was proposed and built were rejected by their citizens. The BRT can be made more sustainable if the switch can be made from diesel buses to electric ones. An article in study quoted in The Scroll (2016) contends that an electric bus emits about 50 kg less carbon dioxide per day than a diesel bus. Presently, the BRT system in Amritsar uses diesel-fueled buses. In an interview, the project manager of the BRT in Amritsar explained that the operating cost of electric buses is much higher than diesel ones. He estimated the operating cost of an electric bus at Rs 60 per kilometer as compared to Rs 20 per kilometer of the diesel bus. He further said that the Capex for the e-buses is less, however the operation and maintenance is high. A big challenge for cities is to get the Operations &Management funds for which the BRT operations need to be profitable.

8.7 Causes of BRT as a Maladapted Solution in Amritsar

Tiwari (2020) argues that the metro is beneficial, only if the source of electricity is a low carbon one, otherwise she contends that a CNG bus is much better than the metro which runs on electricity from fossil fuels. There are several reasons why BRT can be called a maladaptive strategy for climate change mitigation as listed below.

8.7.1 The Felling of Trees

The environmentalists resisted the BRT programme since thousands of trees were cut down to make way for this project. The Hindustan Times (July 2014) reported that there were 1,187 trees which were coming in the way of this project. Of these, 802 were to be cut whereas 385 were to be saved. No public consultations were conducted during the planning of the BRTS project. This led to anger among the citizens. The Project Manager of the BRTS in PMIDC, Chandigarh, however said that more trees had been replanted near Verka village as compensation for this.

8.7.2 Incomplete Network

The BRTS network is incomplete and hence not a preferred mode for many passengers. A network approach instead of a corridor one can make the BRTS more functional in Amritsar.

8.7.3 The Competition with Auto Rickshaws

The BRT runs on limited routes and doesn't connect to all parts of the city. For instance, it is not connected to the old city which is the prime tourist destination in Amritsar. It has not been efficiently linked to the IPT which has a high modal share in Amritsar. Hence, it falls short of meeting the requirements of both the local population and the tourists.

Moreover, the auto rickshaws compete with the BRT and become the preferable mode of transport due to the last mile connectivity they offer. The project has been a contentious one ever since its inception. The auto-rickshaw drivers' union opposed it on grounds that the route was proposed on the roads which were pre-established auto routes. The project would not provide last mile connectivity and would only create chaos and competition on the roads.

8.8 The Use of Diesel-Fueled Buses

The experts believe that diesel-fueled buses in cities should be replaced with electric ones. Or if that is not an immediate possibility due to their high operation costs then the existing buses should be retrofitted as hybrid ones to increase their fuel efficiency.

8.9 Traffic Congestion and Shortage of Fleet

In the survey, the residents of Amritsar conceded that the city needed a fleet of minibuses to ply on its roads and not the big buses that had been purchased under the project. The bigger buses had created more congestion on the streets. Yet others complained that the bus service was erratic, and the buses often crowded. While the respondents of the survey had mixed opinions on whether the BRT had resolved some of the traffic woes in Amritsar, they were unanimous in demanding the BRT on more routes than the currently operational ones. Many respondents said that the BRT needed to connect residents of Ranjeet Avenue and airport to different parts of the city. Transport experts have iterated the need to run pan-city for it to be successful.

Some people opined that making the BRT corridor in the middle of the road had caused wastage of road space and made it more accident-prone. The Project Manager of the BRT in the PMIDC however explained that constructing the BRT bus shelters and lanes in the middle of the road had hugely cut the cost of the project than appropriating both sides of the road for the construction of the same.

8.10 Spatial Strategies to Make the BRTS More Adaptive in Amritsar

We propose a few strategies to make BRTs more efficient and climate-friendly.

(i) Expand the network

The people of Amritsar said that the BRT would be more usable and useful if it connected to different parts of the city. They recommended that the airport line should be made operational soon. Additionally, Batala Road, Ajnala Road and Majitha Road should also have a BRT network connecting to the core of the city. Many residents expressed that the BRT should be operational on the circular road too, yet others said that it should cover all major bypasses and entry points to the city.

The Smart Cities Council (2016) explains that in "Ahmedabad the entire BRTS corridor was designed by following the ideology of connecting busy places but avoiding busy roads. This ideology played a strong part in how the first corridor was selected for design and implementation. The other major point which planners of Janmarg had in mind was that they knew they were 'designing a network, and not a corridor.' The mistake that many other cities have made while visualizing similar projects is that they have thought of them in terms of corridors rather than a network."

(ii) Provide parking facilities near major BRT stops

Like Ahmedabad, Amritsar too should provide parking facilities along the BRT network, especially along the new routes that are going to be developed. Parking may not be a possibility in the core of the city, which is already quite congested but vacant land parcels could be available along the airport road, Ajnala Road, Jalandhar Road, Majitha Road and Tarn Taran Roads (see Parashar, 2014).

(iii) Develop a trunk and feeder system integrating the BRT and the auto rickshaws

On the current routes of the BRT (Phase 1) auto rickshaws are competing with the buses. The administration should regulate the number of auto rickshaws operating on the BRT routes. Auto rickshaws should complement the bus system rather than compete with them. A mobility zoning can be done in the city where the auto rickshaws are the dominant mode in the inner core of the city while the buses should connect the city to its peripheries.

8.11 The RAAHI Scheme: From Diesel to E-Auto Rickshaws

The RAAHI scheme was introduced in Amritsar under the City Investments to Innovate, Integrate and Sustain scheme (CITIIS). It is the main component of the program to fund Smart City projects that were launched by the Ministry of Housing and Urban

Affairs (MoHUA) on 9th July 2018. This Program is financed by the French Development Agency (AFD) and supported by the European Union (EU). The Program is coordinated and managed by the Program Management Unit (PMU) set up at National Institute of Urban Affairs (NIUA), New Delhi.

According to the Detailed Project Report of the scheme, its main feature is to address the pollution caused by diesel-fueled auto rickshaws by replacing them with e-auto rickshaws. The objectives of this scheme were as follows:

1. To Replace 7500 diesel auto rickshaws and provide modern electric vehicles to Auto Rickshaw drivers under Phase 1.
2. Provide better mobility options and improved accessibility to PT and IPT services for citizens and tourists.
3. Provide effective Feeder and complimentary services to the Metrobus corridors.
4. Provide better livelihood opportunities for auto drivers and their families.
5. Improve urban environment by reducing air and noise pollution.
6. Focus on formalization of the IPT sector through formation of Co-Operative Societies.
7. Ensure improved inter-departmental coordination between key Government stakeholders like Municipal corporation of Amritsar (MCA), Amritsar Metrobus, Regional Transportation Authority, Amritsar (RTA) & Punjab Police, Traffic-wing.
8. Provide a comprehensive system of charging infrastructure.

According to the feasibility report of the BRT (2015), there were more than 18,000 auto rickshaws in the city in the year 2014. In several camps organized under the RAAHI scheme in July 2023, a total of 300 applications were received from diesel auto drivers to make the switch to the E-auto rickshaw (data obtained from RAAHI project team).

The E-auto rickshaws are a "winning solution" say experts since they aid in reducing noise and air pollution. If compressed natural gas autos are replaced with e-rickshaws, at least 1,036.6 tons of CO_2 emissions per day (378,357 tons CO_2 yearly) may be reduced. They produce far less noise pollution, making them a more comfortable and pleasant mode of transportation for everyone. This is especially important in urban areas where noise pollution can be a significant issue.

The government is giving several incentives to switch to E-auto rickshaws in different states. For example, the Tamil Nadu government has a clearly defined e-vehicle policy which wants to develop Tamil Nadu as an e-vehicle manufacturing hub, accelerate the adoption of e-vehicles, enhance the development of EV ecosystem in India and develop EV cities in India. The charging stations envisaged in the policy are powered by solar panels and wind energy. It offers subsidies for making the switch to electric vehicles and has made huge advances in switching over to renewable energy. Tamil Nadu has 16 GW installed capacity of renewable energy.

According to Punjab State Power Corporation Limited (PSPCL), thermal power plants are not able to generate power to their full capacity (India Today, 2022). Punjab is currently getting 4336 MW of power, including 1145 MW from state-owned thermal power plants, 2680 MW from private thermal power plants, 358 MW

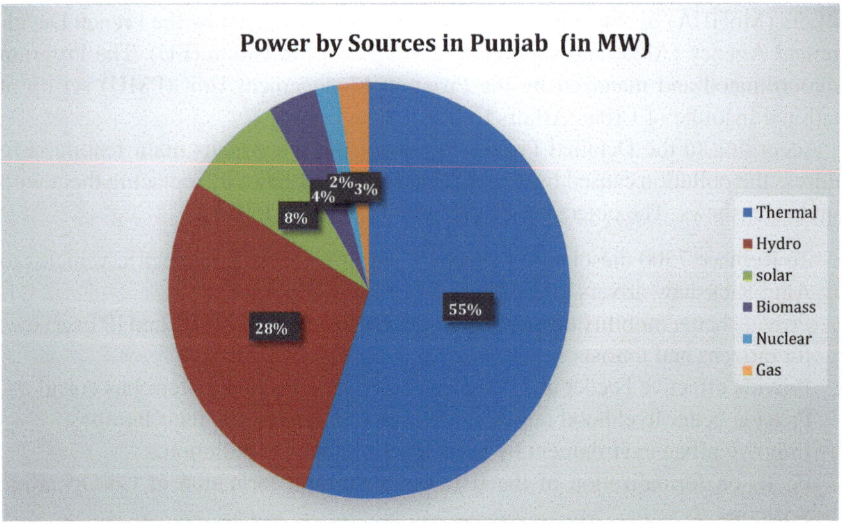

Fig. 8.4 Power sources in Punjab. *Source* Adapted from Singh et al. (2020)

from hydropower plants, besides 153 MW from other sources. The switch to EVs makes sense if the state in question produces its electricity from renewable sources of energy like sun, hydro and wind. But Punjab still heavily relies on fossil fuel-based energy. This means more than 60% of the power in Punjab is generated using thermal energy (coal-based). For the E-vehicles schemes to be "cleaner" or effective, the source of electricity should be a renewable one. However, in Punjab this is not the case. The charging infrastructure that is being developed should be solar or wind powered (Fig. 8.4).

8.12 Strategies to Make the E-auto Rickshaw More Sustainable

From the analysis of our study, we could recommend a few strategies that could make the electricity-auto rickshaws more sustainable. They are as follows:

(i) *Give subsidies to all auto drivers that register for the switch*

The current feature of the scheme is that the first 7500 auto drivers who opt for the switch are eligible for the subsidy. This feature should be amended to cover all the drivers who register for this scheme. Electric-auto rickshaws are more expensive as compared to diesel ones and the auto drivers interviewed for this study expressed that they needed support to buy the E-auto rickshaw.

(ii) *Develop green charging infrastructure and stations*

Like Tamil Nadu, the charging infrastructure should be powered by solar energy (Tamil Nadu Electric Vehicle Policy, 2023). Currently the Amritsar city administration has floated tenders to develop charging stations. In the meanwhile, the charging infrastructure has been provided in the homes of the auto drivers. Solar panels should be installed in the homes to generate green energy to charge the batteries of the e-vehicles.

(iii) *Operationalize the green corridors and zones mandated in the Punjab EV policy*

The most polluted parts of the city should be green-zoned as per the Punjab Electric Vehicle Policy (2022). This should include the walled city which has a heavy footfall of tourists. Only E-auto rickshaws and cars should be allowed in the walled city.

(iv) *Establish Battery service centers*

E-battery maintenance and service centers should be set up in different parts of the city.

8.13 Discussion and Conclusions

The analysis of the two schemes has demonstrated that although these schemes are big steps toward improving the air quality in the city of Amritsar, they can be made better adapted toward climate change by further interventions. These schemes are presently partially contributing to climate change mitigation but could further lead to climate change mitigation if properly implemented. Further, they are presently not integrated with each other but rather operate piecemeal. These two modes of transportation should be fully integrated with each other in a trunk and feeder system on the lines of the Ahmedabad Janmarg model.

The BRTS is currently limited in its outreach and does not cater to the old city and other key routes deterring people from using it. On the other hand, auto rickshaws are the lifeline of the old city which is congested and has narrow streets. A transport zoning of Amritsar should be done on the lines of the dominant mode of transport. Additionally, the green zoning and green corridors of the E-vehicle policy in Punjab should be implemented in the city on a priority basis.

Further, the existing fleet of buses runs on diesel. These should be retrofitted to be hybrid—utilizing electricity and diesel to improve their performance. The Government of Punjab can learn from the Government of Tamil Nadu to introduce battery replacement schemes for e-vehicles in the cities. The charging infrastructure should strictly be powered by renewable energy sources. Solar panels should be installed at all charging stations and homes of auto-drivers. We argue that charging infrastructure should be publicly as well as privately provided across the city of Amritsar.

The competition of auto rickshaws with the BRT is another major limitation that is not addressed adequately by the local Government of Amritsar. The BRT tickets are much cheaper than the auto rickshaw fare. However, the perception survey results revealed that people find the buses overcrowded and infrequent during peak hours and hence rely upon auto rickshaws or private vehicles. To increase usage, the administration should promote the BRT through publicity campaigns and address the issue of inadequate frequency by increasing the fleet of buses. Auto-rickshaws should not be allowed on the main trunk lines and be limited as feeder service to the BRT.

Apart from these local scenarios, the national urban transport policy needs to favor the Bus Rapid Transport Systems over metro rails, which are more expensive and polluting as proven by Tiwari (2020) in her analysis. The shift in the Ministry of Housing and Urban Affairs' focus away from the BRT to promoting metro train systems even in the smaller towns is not a feasible option given its high costs and emission levels.

Indian cities have been rejecting the BRTS, especially the ones in the North due to their reliance on private vehicles, but the fact remains that not all cities have the adequate population size to make the metro rail feasible. City governments are increasingly pushing for metro rail systems at the behest of central government preferences.

The roads in these cities cannot be expanded to accommodate the ever-increasing fleet of private vehicles and people will have to switch to public transport in the coming years. The government should discourage the use of private vehicles by imposing high taxes on car buyers, removing diesel autos from BRT main routes and integrating intermediate transport with mass rapid transport (the BRTS). Additionally, the fleet of BRTS buses should be upgraded regularly and electric buses should replace diesel-fueled buses in cities. The size of the buses purchased under the scheme should be according to the road widths in the city. In Amritsar, many people complained about the big size of the buses which were creating more congestion on the roads. In the perception survey several people opined that minibuses should have been purchased for Amritsar.

The RAAHI scheme is a well conceptualized scheme that will reduce the air and noise pollution in Amritsar. The main limitation presently is that the vehicles are charged using electricity generated from coal-based thermal power plants. We argue that the charging stations should completely be powered by renewable sources of energy such as solar and wind. Further, the scheme limits the number of diesel auto rickshaws eligible for subsidies to 7500. We argue that all auto drivers who want to switch to cleaner fuels should be subsidized. Further battery swapping facilities and service centers should be set up at accessible points in the city.

Presently politics and lobbying by the auto-rickshaw associations leads to them being competitors of the BRTS. For example, the Times of India (2018) reported that the auto-drivers were opposing the BRT in Amritsar. The administration of the city needs to step up the measures to make the auto-drivers aware of the benefits of the mass public transport system in the city and motivate them to operate as feeder services to the main BRT trunk lines rather than competitors.

A decongestion strategy for the city should be based on mobility zoning—with exclusive zones for pedestrians, IPTs and for the mass transit public transportation. These different modes should be integrated with each other to make the city's mobility plan functional. Further, the emphasis on switching to E-vehicles should be complemented with measures like generating power from renewable sources, developing green charging infrastructure and setting up battery exchange and recycling units.

Acknowledgements The author would like to thank the French National Research Institute for Sustainable Development (IRD, France) for funding a part of the fieldwork for this research. The views of the authors are personal.

References

Babalik-Sutcliffe E, Cengiz EC (2015) Bus rapid transit system in Istanbul: a success story or flawed planning decision? Transp Rev 35(6):792–813

Comprehensive Mobility Plan (CMP) of Amritsar (2012) https://dokumen.tips/documents/comprehensive-mobility-plan-for-amritsar-city.html. Accessed 19 July 2023

Feasibility Report Rejuvenation of Auto Rickshaw Drivers in Amritsar through Holistic Interventions (RAAHI) (2020) under CITIIS Programme, A report prepared by the Project Management Consultant for Amritsar Smart City Limited. Document number ASCL/CITIIS/Feasibility Report/12/2020/Final

GoI (2014) BRTS Projects sanctioned across 11 cities in seven States under JnNURM. https://pib.gov.in/newsite/PrintRelease.aspx?relid=103639. Accessed 20 June 2023

Hidalgo D, Pereira L, Estupiñán N, Jiménez PL (2013) TransMilenio BRT system in Bogota, high performance and positive impact–Main results of an ex-post evaluation. Res Transp Econ 39(1):133–138

India Today (2022) Punjab power crisis: thermal power plants lack adequate coal supplies, March 30. https://www.indiatoday.in/india/story/punjab-power-crisis-thermal-power-plants-lack-adequate-coal-supplies-1931189-2022-03-29. Accessed 20 May 2023

Jaiswal A, Dhote K, Krishnan R, Jain D (2012) Bus rapid transit system: a milestone for sustainable transport: a case study of Janmarg BRTs, Ahmedabad, India. OIDA Int J Sustain Dev 4(11):45–62

Jha R (2020) Have Indian cities bid farewell to the Bus Rapid Transit System?, Observer Research Foundation, https://www.orfonline.org/expert-speak/have-indian-cities-bid-farewell-bus-rapid-transit-system/. Accessed 19 June 2023

Kolstø DB (2017) Amaravati: speculation and uncertainty in the new capital city of Andhra Pradesh, India, Dissertation, University of Oslo

Parashar L (2014) BRTS, Amritsar, Punjab, Asia BRT conference, https://www.scribd.com/document/303372057/2b-2-BRT-Amritsar-LaghuParashar. Accessed 30 June 2023

Punjab Electric Vehicle Policy (2022) http://olps.punjabtransport.org/Punjab%20Electric%20Vehicle%20Policy%20-%202022.pdf. Accessed 14 July 2023

Rana R (2022) How could Ahmedabad make BRTS a success story while other cities could not? https://thelogicalindian.com/uplifting/ahmedabad-brts-bus-transport-35111. Accessed 13 May 2023

Sampat P (2016) Dholera: the emperor's new city. Econ Pol Wkly 23:59–67

Singh B, Szamosi Z, Siménfalvi Z, Rosas-Casals M (2020) Decentralized biomass for biogas production. evaluation and potential assessment in Punjab (India). Energy Rep 6:1702–1714

Smart Cities Council (2016) How Ahmedabad succeeded in the BRTS. https://www.smartcitiescouncil.com/article/how-ahmedabad-succeeded-brts. Accessed 23 May 2023

Tamil Nadu Electric Vehicle Policy. https://investingintamilnadu.com/DIGIGOV/StaticAttachment?AttachmentFileName=/pdf/poli_noti/TN_Electric_Vehicles_Policy_2023.pdf. Accessed 20 May 2023

The Hindustan Times (2014) Rs 600-crore BRTS project kicks off in Amritsar, July. https://www.hindustantimes.com/punjab/rs-600-crore-brts-project-kicks-off-in-amritsar/story-qekn880sE3xUaxsz5DlwJP.html. Accessed 30 May 2023

The Times of India (2018) Amritsar BRTS to be revived soon. https://timesofindia.indiatimes.com/city/amritsar/amritsar-brts-to-be-revivedsoon/articleshow/62386959.cms. Accessed 20 May 2023

The Tribune (2021) Resume metro bus from Gheo Mandi to airport, Amritsar residents, September. https://www.tribuneindia.com/news/amritsar/resume-metro-bus-from-gheo-mandi-to-airport-say-amritsar-residents-307717. Accessed 30 May 2023

The Tribune (2023) Review BRTS in Amritsar: Kunwar Vijay Pratap Singh to govt, March. https://www.tribuneindia.com/news/amritsar/review-brts-in-amritsar-kunwar-vijay-pratap-singh-to-govt-486865. Accessed 3 June 2023

The Scroll (2016) Electric buses earn 82% more profit than diesel daily, March 22. Electric buses earn 82% more profit than diesel daily. Accessed 28 May 2023

Tiwari G (2020) Climate change and urban transport in India—You Tube, January 9. https://www.youtube.com/watch?v=Mj_Q79jrJIM&t=129s. Accessed 27 May 2023

Chapter 9
Mountains Are Calling, for Help: An Anthropological Analysis of Tourism-Induced Maladaptation

Kamal Choudhary

Abstract The recent incidence of land sinking in Joshimath town of Uttarakhand, India, had created an alert among the natives. There are many regions in the state that could meet the same fate in the near future if the level of anthropogenic activities is not brought under control. Of these, the most important and often neglected area is tourism. The young fold mountains of Himalayas have attracted tourists because of the peaceful environment. However, over the years the multiple facets through which tourism operates in these hills have done worse than good for the local economy and climate. Of these multiple facets, the following study will deal with the issue of second home and retirement home-induced pressure on the hills of Uttarakhand. Post Covid-19 pandemic, the ephemeral nature of human life and 'live in the moment' philosophy has developed a culture of *leisure-lull* in the non-urban socio–cultural milieu where the urban dwellers could take refuge from the hectic and draining urban life. Unfortunately, this culture is more maladaptive in its practice and has damaged the environment in the name of personal healing. The following attempt is a mixed approach: rooted in empirical investigation via anthropological methods of data collection like observation and interviews; and seeks help from already existing literature on the subject to develop an in-depth and diachronic understanding around the phenomenon. Therefore, efforts are made to explore the phenomenon in depth so that it will be practiced in a more sustainable way by future aspirants.

Keywords Maladaptation · Second home · Urbanization · Uttarakhand · Hill urbanism · Casual urbanism

---

K. Choudhary (✉)
Department of Anthropology, University of Delhi, Delhi, India
e-mail: kchoudhary@anthro.du.ac.in

A. Sarkar et al. (eds.), *Risk, Uncertainty and Maladaptation to Climate Change*, Disaster Risk Reduction, https://doi.org/10.1007/978-981-99-9474-8_9

9.1 Introduction

The ancient Greek myth of Ouroboros (parallels found in Egyptian and Vedic texts as well) where a serpent is represented eating its own tail is a symbolic depiction of an eternal cycle of destruction and re-creation or of birth, death, and rebirth. This cycle in many traditions is considered a trap for the human soul and reason for all suffering. While the cycle is itself inescapable (if not undertaken any religious or spiritual escape), human deeds have obviously been found responsible for speeding up the process of destruction and hence, further getting entrapped in suffering. In the context of this chapter, human adversities rooted in maladaptive practices, the myth fits well. Human behavior is majorly dictated by cultural practices through which they learn to engage with other fellow beings and nature; as long as these practices are drawn by exploitative economic and individualistic gains, humans will continue to suffer by exposing them to social and ecological vulnerabilities.

On a global level, there has been a demographic shift where the majority of the population are now residents of urban areas which is expected to rise up to 68% from the current 55% by 2050 as stated by UN Department of Economic and Social Affairs (UNDESA 2019) report 2018 Revision of World Urbanization Prospect. While the same report opens up with a daunting statement: "The future of the world's population is urban" (UNDESA 2019), the human spirit says otherwise. The two prime reasons for this shift are: firstly, the urbanization of rural areas whether voluntary & planned or forced & unplanned, via processes like peri-urbanization and suburbanization; secondly, the people migrating from non-urban areas into cityscapes in search of a financially stable life. It really does not matter how one journeyed into the urbanization process i.e., whether one is forced into it, or one has been the catalyst to this process, both will eventually end up having a saturation of urban life. The catch is while the former who are forced into this change already had an idea of rural life and hence have nostalgic elements that could delay the saturation process for a while, but for the latter who have been habituated (Bourdieu 1977) into urban lifestyle, this threshold is far easy to achieve especially when they are surrounded with an escape route provided by mountains and coastal tourism with all its idyllic representations and cliches funneled down by media and gossips. Hence, for an urban dweller whose lifestyle patterns are defined by conspicuous consumption (Charles et al. 2009; Heffetz 2011, 2012; Veblen 1899) and one that enjoys the privilege of affordability (Osbaldiston 2012) and mobility (Bloch 2020; Sheller and Urry 2006), taking an escape from the draining urban life becomes initially a need and a lifestyle eventually. Therefore, this way of life gives rise to a new form of urbanism in the hills rooted in 'live in the moment' philosophy, ending up consuming its space, place, and resources in a way different from the way local/natives use them, which is sometimes on the more casual or far away from sustainable practices.

The situatedness to this *casual urbanism* of the hills can be found in the age-old phenomenon of second homes. Second home, though seemed to be a contemporary phenomenon because of its growing abundance, has a long historical rooting that can be traced back to Egyptian and Roman times. The migration from one's primary residence to a different location in search of spatio-temporality which is different in its *feel* (Osbaldiston 2012), its *rhythm* (Parkins and Craig 2006), *vibe* or any other neologism that could exist just in order to differentiate or create a binary between primary home and the new location is an important prerequisite while dealing with this phenomenon. The primary motivation for second investments is rejuvenation of mind and body (this Cartesian dualism is a result of how owners of second home explain their experience of the house), apart from other spurs like low-housing rates, affordable retirement life, commercial holiday homes, etc., that will be dealt in depth in later part of this chapter.

When it comes to mapping of the regions that are receiving second home investments, the geographies vary across the domain of Sea-change, Tree-change, Green-change to Hill-change (Osbaldiston 2012) and include regions of Sweden (Müller 1999), Poland (Adamaik 2016), Africa (Hoogendoorn and Visser 2004; Visser 2003), France (Hoggart 1997), Spain (Barke 1988), China (Wu et al. 2015) and Australia (Osbaldiston 2012) and others whose picturesque landscape can provide the onlooker with calm. While the intensity of second home investment in these regions may vary, it is sufficient not to go unnoticed especially in the way it has brought socio-cultural and topographic changes to the landscape; becoming a pan world phenomenon receiving investments from both domestic and international tourists. Hence, this chapter is an attempt to bring to fore the development of second homes in India and how some of unsustainable and maladaptive practices of its owner and those related to it, in one way or other, are bringing a bad light to otherwise this value neutral phenomenon of human dwelling.

9.2 Knocking the Door: Methodological Approach Toward Second Home Study

While embarking on a journey to understand the enigmatic phenomenon called 'second home,' I had my personal selfishness. Being an urban dweller, raised amidst the turmoil of city life, the mere sight of the majestic mountain peaks ignites a profound sense of tranquility with a simultaneous aspiration to envision a home of my own nestled amidst nature's lap. However, the disparity between this idyllic vision and the harsh ground realities has left my 'mountain person' in a state of apprehension. While I refuse to choose to become another urban dweller contributing to the scars left on the pristine landscape, I also could not let go off the dream of mine and countless other kindred souls whose heart beats for the serenity of mountain homes. Hence this pursuit is to find a delicate balance—an alchemy that transf orms aspirations into responsible actions, dreams into sustainable realities.

While within India, there are some great landscapes that are on radar of second home investors which range from beautiful beaches of Goa to picturesque tea gardens of Ooty in Tamil Nadu and obviously lush green meadows of Mahabaleshwar of Maharashtra and Coorg of Karnataka, but Uttarakhand state is a perfect tapestry

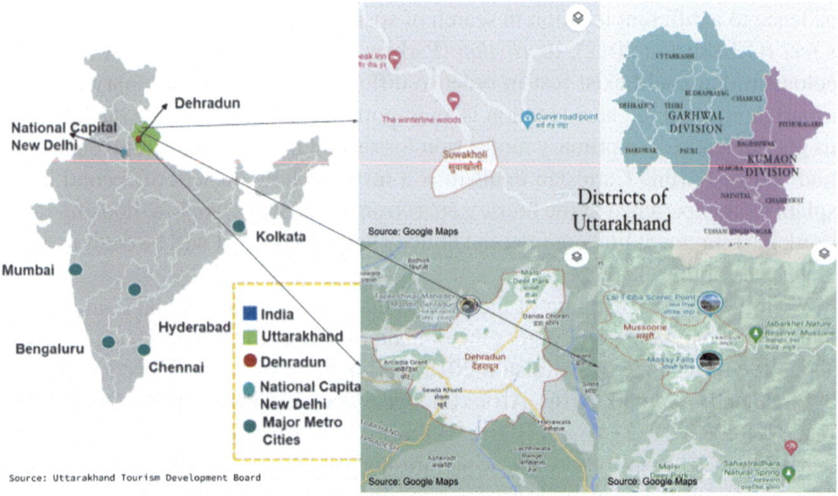

Fig. 9.1 Map of Uttarakhand, along with three sites of fieldwork

to start the investigation. Uttarakhand receives a huge diversity of young and old tourists who crave for adrenaline rush through adventures and sports offered by Mukteshwar, Auli and Nainital while others who seeks spiritual refuge in Haridwar, Rishikesh, Kedarnath and Roorkee (Nakano et al. 2017). The liberal and alluring land buying laws for outsiders[1] have also made Uttarakhand as a perfect destination for second homeowners. This chapter includes case studies and narratives of second homeowners, the local brokers who mediated the land deal and the locals/ natives who witnessed this change from the towns of Dehradun, Mussoorie and Suwakholi (Fig. 9.1).

While anthropological fieldworks are usually long-term engagement for almost a year that produces ethnographies which are holistic at least in its seasonality, but for the current study, I took refuge of what Renato Rosaldo called "deep hanging out" (Clifford 1997: 351) which helped to follow the mobile residents of a rooted dwelling structure. Moreover, since second home investments are not very dense at present to form a geography-specific community, the best way out is to trace the phenomenon at different places, giving the study a touch of multi-sited ethnography. In total, 42 second homeowners and aspirants, 15 local residents, 7 administrative officials and 12 real estate agents have been engaged in in-depth interviews. Apart from this, secondary data based on land records from Tehsil office, local courts and survey office have been collected to supplement primary data.

[1] The term 'outsider' has been used throughout the chapter for the non-residents of the state who could not claim any ancestral connection to the land. Also, this term is used by the locals/ natives themselves. It is important to clear this because the label of 'being an outsider' is highly political especially for towns like Dehradun whose major residents today are outsiders of the past.

9.3 Conceptualizing Second Home Investments

9.3.1 *Definition and Characteristics*

Presently the scholarships under which second home studies can be put are of varied degrees in the way how the process is understood. It can be approached as amenity migration as described by geographer Laurence Moss where "environmental" and "cultural" (Moss 2006: 8–9) prospects drive this form of tourism; it can be approached as lifestyle migration where the entire movement is explained via fixed and patterned choices dictated by habitus (Benson and O'Reilly 2009); or it can be explained via consumption-led migration (Visser 2003; Hall and Müller 2004) where focus is on how the second home owners consume the landscape and induce changes at the same time. While there are many parallels in these approaches, hence not as distinct as one believes them to be, they forced the creation of varying definitions of a second home. The sine qua non for differentiating a second home was that its owners have to have a permanent residence (Ragatz 1977; Tombaugh 1970) and the 'second' in second home is associated with its temporary-ness as the owners have to return back to their primary abode from time to time, but this is also challenged by scholars who believe that the emotional capacity/attachment of second home often surpasses that of primary residence to an extent that they should be rather called as *first home* (Kalternborn 1998). Hence, the variation in second home conceptualization appears in terms of their utility (re-creation, leisure, retirement, work), occupancy (bachelors, nuclear family, joint families, visiting friends and relatives) and size (cabins, cottages, apartments, villas, farmhouses, etc.). Thus, to create boundaries around this otherwise variable and individualistic phenomenon, the present study adopted the definition offered by Goodall (1987) which says second home is "a property owned or rented on a long lease, as the occasional residence of a household that usually lives elsewhere" (in Hoogendoorn and Visser 2004: 107).

Within second homes, there are further categorizations based on the motivation of the second homeowners or investors. These motivations further dictate the variables like ownership, utility, mobility and scale on which dwelling in a second home is dependent. For this, Wu et al. (2015) produced a criterion based on their study on the second home city of Sanya in China where they made the distinction on the basis of property ownership, use of second home and mobility pattern. The resulting dimensions are:

- Mobility pattern: this dimension decides the span of time spent in a second home and can be categorized as *weekend vacation, short-term holiday, seasonal migration and permanent migration.*
- Property ownership: this dimension takes into consideration the extent to which second homeowners pump capital into properties based on its usage and available mobility pattern. It includes *private house, intermittently private house, intermittently commercial house and commercial house.*

- Use of second home: this dimension is a result of mobility pattern and ownership which finally decides how the second homes are actually utilized. If a house is used on weekends or short-term holidays and is privately owned, they are categorized as *elite-vacation homes* like luxury villas. Similarly, *lifestyle migration homes* are "houses rented on a long-term lease or purchased for personal or family use. The age group for this owner category spans a wide range, including young families and pre-retirees who may not be able to afford the higher end luxury houses" (Wu et al. 2015: 145). Finally, *retirement-migration homes* are aged communities with small but privately owned property.

This categorization is helpful not just in mapping the motivation of migrants to invest in a second home but also in analyzing the socio-spatial relation it generates with local ecology. Often the owners of *elite-vacation homes* and *lifestyle migration homes* who are short-span visitors have a preference for secluded life to an extent that they create physical and abstract boundaries of interaction from the locals/natives. In Suwakholi, almost all the second homes have such boundaries which in some cases go as explicit that they are further covered by translucent to opaque sheets (as shown in Fig. 9.2) to minimize the visibility of inner world to outer world and outer world to inner world. While some may argue this as an act toward privacy and security of residents but a geography which is popularized (by natives and media discourse) and imagined (by the second home investors) for its "simple and peaceful living" as against the crime brooding cities (these binaries are key while marketing hills landscape as against the metropolis as shown by Osbaldiston 2012), however does not quite fit this type of seclusion with that argument. Hence, as a result such types of dwelling of habituated urban dwellers produce urbanism which is far from the local way of consumption. The gated societies in Dehradun whose apartments are mostly owned by residents of Delhi and other North India city residents are furnished with lavish facilities like swimming pools and golf courses when the state suffers from an intense problem of water scarcity.

While, on the contrary, the *retirement-migration home*owners who eventually end up settling in the place are far more connected. It is believed that "They actively participate in public activities, they organize cultural affairs for the community, and they concomitantly work to bridge the communication gap between local residents and other seasonal migrants. They are, however, no longer second-home tourists, though they might not be authentic locals, either" (Wu et al. 2015: 150).

Hence, second home development in a mobile world is a result of experience economy (Pine and Gilmore 1999), if seen entirely from economic sense where the experience of living in a second home comes with pre-generated imaginations and packaged deals, especially while buying an apartment or condos from an already existing real estate company whose sole aim is to capitalize on this emerging phenomenon.

Fig. 9.2 Image showing second home covered by steel mesh layers and further covered by translucent sheets. *Photo* By author

9.3.2 Growth in Second Home Ownership and Driving Factors

In context of second home development in Uttarakhand state of India, the statistical data is zero or at least not out there on public platforms. The major setback is the unavailability of any survey to track the accumulation of property acquired by an individual or household outside their current residence and resident state. Hence, deciphering anything statistically on a second home is difficult. However, according to the statistical compendium of State of Housing in India 2012, with a national deficit of 18. 78 million houses, the state of Uttarakhand alone accounts for 0.16 million house shortage. This issue has been brought forth by the scholars who, along with this aspect, have produced findings that second home investments had created a negative effect for the local community in terms of lack of housing facilities and increased property rates (Müller 2002; Gallent et al. 2005; Gallent 2007). This study, although could not contribute explicitly in terms of numbers to this debate of second home investments, rendering the locals/natives without a roof, but there is enough evidence to show that land rates have been skyrocketing ever since the elites of urban areas have set their eyes on this region. The circle rate[2] in Uttarakhand has undergone revision in 2023 and the rates of some areas of Dehradun have witnessed a hike of 150% (Singh 2023). The differential affordability of the city dweller versus that of a hill resident is the key factor that contributes to this price rise because "amenity migrants are broadly speaking quite well resourced in comparison to those who live in small country towns and villages. To suggest otherwise underestimates the income and wealth disparity that exists between city and country populations" (Osbaldiston 2012: 4).

While studying a second home has been historically understood as an elite phenomenon where affordability is the main concern, recently a number of factors have contributed to its development, bringing it out of its mere economic interpretation. These factors include:

Custom fit: As opposed to the affordability concern of second homes, the key real estate players have gone beyond, to target a much bigger audience of youth who in their young age do not have a type of savings or earnings that could be utilized for owning a home. Hence as a substitute, they provide alternatives that go easy on pocket and yet do not compromise on the *feel* that one came looking for in the mountains. The emerging players in these sectors include Airbnb, Zostel, Go Stop, etc. This aspect also forced epistemological changes starting with the very definition of second home in terms of its ownership and family-laden outlook.

Globalization and Neoliberalism: As the economic investments have surpassed the national boundaries and individuals are able to easily invest in a foreign nation, many from abroad are investing in countries where the capital (economic, social and cultural) exchange rates are beneficial. With ease of mobility and cosmopolitanizing of taste (Sheller and Urry 2006), the performance for exotic geographies and authentic cultures has become less burdening. Hence, making the second home dwellers to actually enjoy their time off from city life instead of taking tours like an over enthusiastic tourist seeking exoticness of the space and place.

Technological advancement: Devices like mobile phones have transformed the travel landscape, influencing both the activities of the travelers and their interaction with the destination. But in the context of second home development, the real technological contribution came with ease in communication via virtual meeting platforms and better internet connectivity even on mountains that need not require the presence of a physical body. This aspect had benefited particularly the corporate employees for whom 'work from home' (WFH) culture had provided them with the opportunity to witness the same burden of nine to five jobs, but now not in the midst of great urban architectural wonders but in the abode of mountains, and hence giving rise to digital nomads (Makimoto and Manners 1997). Further, the WFH culture with its flexibilities in terms of remote working have blurred the boundaries between work and vacation. Now one can dip their feet in calm and turquoise beach water or enjoy

[2] Circle rate is the minimum or base price of a property. Any property, be it a piece of land or apartment, cannot be sold below this price. These prices are set by the state government and can vary from region to region.

the snow-clad mountains and mint money at the same time from their laptop that could sustain that sort of living.

Carpe diem: This poetic expression of 'live today and worry not about the future' philosophy has been a key driving force for various tourism-related activities. The evidence is more explicit post Covid-19 pandemic when humanity realized not only about its transient nature but also its innate capacity (and privilege) to be mobile, which gave rise to phenomena like revenge tourism (Wang and Xia 2021). Therefore, the movement from survivalist elements like capital accumulation to self-expression which calls for emancipative ethos gave rise to a behavior where people do not hesitate to invest in their dreams, aspirations and fetishes.

Hence, as the dimensions of second homes have grown beyond vacation homes the resultant investments in this sector are huge. People are buying individual second homes or simply building commercial spaces that could be utilized by others for their urban escape (Osbaldiston 2012). Therefore, either for self-consumption or for capital generation, investment into second homes is a win–win deal. Due to this very fact, the current situation in Dehradun and nearby towns is explained by a local property dealer, Madan[3], who says "Most of the land that could be sold, is sold. The major portion of the land adjoining the roads in Dehradun, Mussoorie, Suwakholi and other places either belong to *Dilli-wala* (Delhi people) or rich of Dehradun." The race of buying land in these hill towns had gone to such an extent that every means was adopted to purchase a "suitable" piece. Madan enlightened this further, "Once a party chooses a land, we see in what category it falls. If it is an *Abadi* land (residential) then there is no issue as an outsider can buy up to 250 m^2, but if it is agricultural land then it is converted first into Abadi and then the deal proceeds. For this, documents are backdated to a time when Pradhan was a key player unlike current Mussoorie Dehradun Development Authority (MDDA) which holds the entire jurisdiction. If someone is interested in land bigger than 250 m^2, which is usually the case, then there are other ways like registration on the name of different family members or giving Power of Attorney to outsiders". These loopholes within the bureaucratic system have paved a way for purchase of land more than prescribed and permissible limits. As this scenario is getting increasingly prominent not just the existing land use pattern is changing (Nakano et al. 2017) but has also enticed the locals/natives to sell their land at a far cheaper rate for short-term financial, social and ecological stability. This aspect is dealt with in depth in later sections.

9.4 Climate Change and Its Implications

In the current epoch of the Anthropocene where human species alone is the most influential in terms of its impact on the planet, it is nowhere wrong to say that climate change is happening, and humans are responsible for such induced and exacerbated changes. But Chakrabarty (2012) problematizes the very term 'Anthropocene' and

[3] All the names used here are pseudonyms to maintain anonymity of the participants.

points out that "If critical commentary on globalization forces on issue of anthropological difference, the scientific literature on global warming thinks of humans as constitutively one– a species, a collectivity whose commitment to fossil-fuel based, energy consuming civilization is now a threat to that civilization itself" (Chakrabarty 2012: 2). The issue with such homogenized accusation of an otherwise highly differential contribution to the phenomena of climate change is that "those who concur with the scientific consensus on climate change recognize both (1) that everyone contributes to climate change and (2) that some people contribute far more than others. But science says nothing about which proposition should be morally emphasized, and that choice matters hugely, making the difference between two starkly different 'moral readings' of climate change" (Rudiak-Gould 2014: 366). Hence, if the accused are some particular sections of humanity, then obviously there is another section which is rendered more vulnerable than others on the same differential scale of social, economic and spatial variabilities. The Global Climate Change Task Force (GCCTF) in association with American Anthropological Association (AAA) produced a report in which they mentioned: "The residents of Coopersville, Maryland once inhabited a flourishing town, boasting the activities of watermen (crab, oyster, and fish harvester), seafood processor, restaurants, and modest vacation homes. With sea level rise, they now face frequent flooding and preparation for the next big storm. In concert with existing concern about rolling easements and dwindling real estate values, the change portended by climate variability and flooding exacerbate an already difficult situation for residents—in this case, climate change intensifies the social and economic declines in that part of the state" (Fiske et al. 2014: 14). Hence, climate change is not a tale from Arabian nights which is a fictitious story of a far-off land but an unfortunate reality of our very own surroundings and needs immediate redressal.

The root cause behind climate change and other human vices is conspicuous consumption, a term given by Veblen, where the distance from a good's "utility" and shift toward its "honorific" aspect had veiled the "waste" the entire process produces in turning an otherwise serviceable product into luxurious item (Veblen 1899). As a result, more individuals desire to acquire luxury and those who already have would consume more to have their status maintained in a structure driven by capitalism. The burden that this entire process puts on earth's natural resources is immense (Mi et al. 2018). Even in the current context of second home-induced investments in the hills the production of these luxurious homes has taken a toll on the mountain ecology and as the number of consumers has increased, especially post Covid-19 pandemic, in order to meet this demand, the production side has gone past the ethical, legal and natural constraints. This consumer culture has blurred the boundaries of needs and desires, and as a result in order to achieve this simulacrum of happiness (Baudrillard 2009; Miller 2013), urban dwellers are ready to get consumed by their tiring jobs which also act as a justification for the "great urban escape" (Osbaldiston 2012). This homogenization of consumption produces affluent cosmopolitans who share similar traits globally (Hannerz 2002) and hence create similar repercussions of over-consumption all over.

In order to understand climate change more closely, it is important to look at it in tandem with the responses generated by humans and for this, these 3 concepts are worth exploring: adaptation, vulnerability and resilience. Adaptation, as understood in anthropology and beyond, is a bio-cultural evolution to stressful situations that ensure survival. Patterson (1994) pointed out that it is through material practices, which are constructed socially and imbued with cultural meaning that humans interact (among themselves and with nature) and sustain. "Humans adapt to their dynamic natural, socio-cultural (including institutional) and built environments through a cultural lens of individually and collectively interpreted knowledge and meaning, to make decisions and respond" (Fiske et al. 2014: 41). However, adaptation is not unidirectional and rather is affected by systematic vulnerabilities in terms of access to resources, which are result of inequalities created in the society and hence the term 'adaptation' evolved from mere change in belief and behavior to even institutional interventions via policy-driven projects and strategies (Nelson et al. 2007; Pelling 2011).

Vulnerability is exposure of humans to environmental, economic, social and political pressures that render them susceptible to hazardous ecological conditions and it is well agreed that social characteristics are prerequisite in order for an individual or community to become vulnerable as it would affect their capacity to anticipate, respond and recover (Wisner et al. 2004). Hence, things gradually get worse for those who are already socially vulnerable as they experience a more frequent and profound impact of climate change. It is therefore important to develop measures that could impart resilient abilities into such vulnerable sections and hence, resilience is defined as reduction in risk and losses to ecological and socioeconomic factors by preparing, mitigating, recovering and adapting to the stressful conditions.

From these three concepts of adaptation, vulnerability and resilience what remains a common point of departure are the cultural practices that not just dictate how one interacts with nature but defines the course of attraction thereafter. If cultural practices are themselves maladaptive, pushing humanity into an abysmal state of living, it becomes very much necessary to get to the roots of these practices to flag them and eventually put a stop to them.

9.4.1 Understanding the Relation Between Culture and Climate Change

There is a dyadic relationship between culture and climate change. While the dimension of climate change affecting culture is well explored (Adger et al. 2013; Kirsch 2001), this study attempts to contribute to the dimension of how culture contributes to climate change. On a behavioral level, people need to consciously realize that the choices they make affect the ecological conditions and hence shifting to more environmentally friendly practices is the need of the hour (Swim et al. 2011). However, there always remains a skepticism over alteration of those habits that are ingrained via cultural upbringing and hence performed unconsciously, which again raises the question of how much agency does one practice over the social structure and to what extent it affects the social structure. While this question is out of the scope of this chapter but needs a thought especially when such maleficent habits are claimed important toward cultural identity.

The meaning-making of physical and metaphysical realities is through culture, people do not use scientific jargons for say climate change but have cultural alternatives which are far much stronger and influential in terms of the impression they create. As the winter kept delaying each year due to prolonged summers and its heating effects, all I remember my mother saying to me every year "You know that in our times it was during Dussehra (festival celebrated in month of Oct-Nov) that we already had our sweaters on and were already snacking on peanuts and *gajak* (sweet made of jaggery)". She was right; I myself felt this change, as our winter clothes were out by the time of Diwali (festival celebrated 20–21 days after Dusshera). This symbolic hinting of changed or delayed sense of dressing and food practice in tandem with seasonal cycles are cultural narratives of climate change. Similarly, our religious and political beliefs affect the way we perceive reality and often these beliefs function as meaning making lenses for complex situations. The gatekeepers of these beliefs have a major role in the way they provide an interpretation of both natural and supernatural processes. The otherwise skeptical Catholics who did not see climate change as an issue soon became active voices after the encyclical of Pope Francis *Laudato Si'* citing climate change as a moral issue and requiring swift collective action (Vincentnathan et al. 2016; Schonfeld and Winter-Levy 2020).

When adaptive strategies are taken into account, countries like Bangladesh have emerged with alternative solutions to its flooding problem each year. The people there have shifted from chicken rearing to duck rearing as the latter can prevent the economic losses caused by flood due to duck's swimming capabilities, but behaviorally also people need to adapt to this changing diet from chicken egg to duck egg (Schipper 2020). Often an adaptive strategy takes no time to become maladaptive if all major and linked factors are not taken into account. Hence, it becomes important to deal with the spectrum of maladaptation by looking at variable causes of vulnerabilities.

In the case of farmers who have shifted to a different subsistence pattern in Uttarakhand, selling off their agricultural land due to poor crop produce of state's climatic and flood conditions, have however not been able to gain the same type of socio-economic security from the short-term capital received. Therefore "this could include a strategy that encourages farmers to sell their land and become employed in another industry that is equally sensitive to climate change impact but gives a wage and in short term offers more security. But it can be maladaptive as it leaves farmers no options to return to farming when the other industry is affected, and jobs are cut" (Schipper 2020: 411). Hence, firstly the locals/natives who sold off their land to property dealers for second home investors have received a sum (approximately 30 lakhs) which is far less from the amount offered to outsiders (approximately 1–2 crores) and secondly, even from the amount received, the locals failed to invest them in any long run venture ensuring security. Usually, they end up building homestays for tourists as even the state government provides subsidies for the same. However, they are not as successful as it is assumed to be because of their non-touristy locations and competitive outside players offering better facilities. As a result, these locals end up becoming the caretakers of the second homes built on their own land or land of someone like them and left complaining "*Hum to apni hi zameen par naukar ban gaye* (We have become servants on our very own land)."

9.4.2 Role of Second Home Investments in Exacerbating Climate Change

In order to understand the relationship between second home investments and induced environmental changes, we have to look at the complex relation between human, nature and consumption. Unlike other factors like vehicular fumes, industrial waste, deforestation, etc., which are directly linked and visible causes of climate change, the changes caused by second homeowners and investors remain inconspicuous to the eyes of those who look for direct links and become visible only in its consequences. Uttarakhand is one of the states of India whose economic output is deeply affected by the tourism industry. The tourism sector alone accounts for 9.68% of the state's GDP and has been successful in generating 17.10% of employment (Saxena and Roy 2022). With each passing year the footfall of tourists increases more than last year. The report of Uttarakhand Tourism Development Board (2019) shows a rise of 105.4 lakh domestic tourists in 2001 to 366.9 lakh in 2018 and 0.54 lakh of foreign tourists in 2001 to 1.56 lakh in 2018. The government is making its utmost efforts to develop this sector to its full potential (while writing this chapter Uttarakhand is witnessing severe floods and parallelly one can find "Invest in Uttarakhand" billboards in Delhi metro), by building roads and bridges, expanding current roads to increase footfall and decrease the time to reach the valley[4], promotion on international and national scale, giving subsidies to local ventures as in the case of homestays[5] and other steps that could fully establish the touristic identity of Uttarakhand. In fact, the impact of this geography has been so profound since colonial times that many of the state's place, place making and the practices (including second home dwellings) have a colonial rooting to it. Hence, Uttarakhand under different regimes has capitalized well on its natural environment and rural landscape.

[4] In mid-2022, after the court's permission to widen Sahastradhara road in order to create a faster route for tourists, Dehradun witnessed axing of more than 2000 trees. The authority paid no attention to activists and local pleas saying that most of the trees were eucalyptus and hence no harm is done rather fast travel time would cut the fuel. (For more details see https://www.nationalheraldindia.com/environment/more-trees-are-felled-in-dehradun-to-make-escape-to-the-hills-faster).

[5] The Uttarakhand Homestay scheme under Deendayal Upadhyaya Grah Awas Yojana was a government effort to improve the economic conditions of the local by providing subsidies over the loan taken for this purpose. With the target to develop 5000 homestays, the bottom line was to promote *Pahadi* culture through food, culture, heritage, and architecture. (For more details see https://www.euttaranchal.com/tourism/uttarakhand-homestay-scheme.php).

As per the definition of second home chosen for current study, it includes varied forms of dwellings ranging from farmhouses, bungalows, condominiums, apartments to rented hotels, homestays and BnBs. Hence, from this perspective the amount of construction that hills of Uttarakhand are receiving in the past two decades is tremendous. While there has been multiple capping over the years on the amount of land that can be bought by the outsiders which kept changing with changing regimes in the state, with the current land ceiling of 250 m^2, yet somehow the outsiders have managed to find the loophole of the social and bureaucratic system when natural resources are commoditized (Ghertner and Lake 2021), and very tactfully managed to not just buy land more than the prescribed limits but also of desired choice, even if it does not fall in the transferable category.

Such a long history of sale and purchase of land in Uttarakhand have given rise to second home dwellers, who either have permanently taken abode in the hill, in the form of retirees or they are urban escapees from work, city or harsh sun, who seek refuge for a while in the midst of mountains. In both cases, a new sort of urbanism is developed by this community which reflects the way in which they interact with local ecology. This urbanism is rooted in 'live in the moment' philosophy and driven by the privileges has developed a culture of leisure-lull or *casual urbanism*, where the second home owners, especially the one that stays for shorter period, are so much into rejuvenation and leisure seeking attitude that they end up casually treating the people, the place and the nature.

The *casual urbanism* is about the value and morals that the outsiders carry with them to these natural landscapes, as their encounter is an experiential product based on responsibilities and social justice (Ericson et al. 2014) or if approached from a more behavioral aspect of tourism, it means a deviation from "empathic Sustainable Development in Tourism (SDT)" (Adongo et al. 2018: 252) where "Empathic SDT is conceptualized as a form of tourism in which stakeholders have strong positive feelings and commitment towards the welfare of local residents, conservation of natural resources and enhancement of tourists' experience" (Adongo et al. 2018). The biggest setback of *casual urbanism* of second homeowners is that in their quest for peace and serenity, they most likely bend toward seclusion where in their imagination of calmness, there is no room for the local community and interaction. As a result, they remain secluded from the local knowledge of utilizing the geography because "members of permanent communities appear relatively well-informed about the risks associated with climate change for their particular locations, for some holding a coherent set of values that can be clearly articulated. A challenge, therefore, for these communities is how to engage the non-permanent residents in ongoing local concerns and issues" (Osbaldiston et al. 2014: 73).

This *casual urbanism* is maladaptive in the way it operates. The worry goes further deep when these short-time dwellers end up becoming long-time natives of the hill geography but with the same urbanized attitude and belief and hence exploiting nature and exposing humanity to future vulnerabilities. The practices encompassing this form of urbanism in the current context of Uttarakhand include:

a. *Excessive and informal way of land utilization*: Uttarakhand since long time has been echoing with demands for more stringent land laws in the state just like its neighboring state of Himachal Pradesh, with locals protesting in front of government institutions raising slogan of "*bhoo kanoon lao, Uttarakhand bachao* (Bring land laws, Save Uttarakhand)". Since the formation of the state in 2000, Uttarakhand has witnessed a series of bouts over land ceiling from 12.50 acres to 500 m^2 to 250 m^2 back and forth. While this issue has always been a hot topic during election campaigns, the results have failed to satisfy the locals. The reason being that the elite urban second homeowners have found ways to buy land of desired size, type and location. It would give some hope, if this exchange would have been labeled as 'under the nose' of the authority but the reality is, it is happening in broad daylight with and without authority's help.

   While talking to a second homeowner and local broker-cum-builder about the issue, they faced while purchasing land, the former jokingly said "*Kuch nahi, Rana ji ne patwari ko bootle de di* (Nothing, Rana ji gave a bottle of liquor to the patwari)." This narrative of rigging has been repetitively observed from the field where the locals pointed out how the fault lies at the very core of the bureaucratic system (Mathur 2016).

b. *Easy money causing uneasiness*: Currently, majority of "aesthetic" land on second homeowners' radar does not lie with the locals but with local elites and builders, but once when it did, locals engaged in agricultural activities sold their land to escape from every year's bad produce due to water stress, flooding and other vulnerabilities to climate variability. Hence, the farmers of high-altitude regions abandoned farming, sold their lands at a price they believed was good and migrated to plain regions of the state. Now in the aftermath, today these farmers look back, they are filled with repentance on two levels; firstly, as they say "*Badi sasti zameen beach di, hume to kuch ni mila* (We sold our lands at very low price and as a result got nothing)". After realizing the potential value of their land today and the amount they could have earned at present versus at which they disposed the land, they count it as a bad deal. Secondly and most importantly, the majority of these farmers have not been able to secure a sustainable and respectable means of livelihood post sale. Through this indication, I do not intend to restrict farmers to their farm economy and disrupt their social mobility but try to point out the level of security land hold on both economic and social scale (Ghertner and Lake 2021). While urbanization seemed to be an inevitable doomsday, the differential benefits made life a living hell.

c. *Unchecked constructions flouting established laws*: The Uttar Pradesh road law and state's by-laws, amended in 2015 and adopted from National Building Code, restricted construction on the sides of the roads and on places where the natural slope is more than 30°; where it is the width of the roads alongside that would determine the height of the building. Both Dehradun and Mussoorie fall under seismic zone IV and such rampant constructions of high-rise projects can repeat the trauma of the 1905 earthquake witnessed by Mussoorie. The recent case of the Sahastradhara tree felling in Dehradun whose supposed rationale was to create a faster route to the hills by increasing its width has an implicit connection.

The Sahastradhara road, connecting to the hill station of Mussoorie, has recently witnessed huge high-rise construction of residential apartments. Further, there are many future projects coming alongside. This sudden step to widen the road can much better be explained by activists' voices who believe "First, all these plans were passed and then the policies were introduced accordingly" (Tyagi 2022).

While interviewing a second homeowner, who had bought a penthouse on the top floor of Pacific Golf Estate in Dehradun, he told "*Jaise hi dusri party milegi, baech denge* (The moment I find an interested buyer, I'll sell it." The penthouse is bought on a price of a whooping 30 million rupees and is not even finished, to say the house has not yet been converted to a home, and its owner is already ready to dispose it off. The reason lies in the fear induced by a mild earthquake in Dehradun a day before this interview which as per the owner had left the entire building swinging. This case points out the priority of any potential second homeowners while buying a piece of land or apartment, of which the seismic bearing capacity of the building or the zone stands last. While the second home aspirants often fall victim to real estate players but are equally responsible themselves for these careless investments, therefore, rendering vulnerable future aspirants of hills to these natural hazards, or should they really be called natural hazards when it is all a result of anthropogenic casual urbanism.

Further, an architect friend based in Dehradun added that before almost 2000 trees on Sahastradhara road were axed, the road had a soothing canopy for a long range; however, at present, the area now receives direct sunlight raising average temperature of the area but also the cars, that were once parked under the canopy get heated in the sun and thus require unnecessary utilization of petrol to cool them. Hence "those who cannot remember the past, are condemned to repeat it", but at the same time it is necessary that these outsiders should be made aware and empathize with this past of which they have never been part of, so that it could induce a sense of fear, responsibility and accountability in them.

d. *Homogenizing place and people*: It is a rare sight to find a second home made from traditional architecture in Uttarakhand. For locals, the traditional style of home is a monetary issue but for this section of the community, affordability is not that big a concern. Rather, what is found is that these second homes are no different from the concrete houses of urban posh localities, installed with the same urban facilities. The same is true with the food practices. In the homestays that I have stayed at and the hotels that I have visited in potential second home areas, it was pretty hard for me to distinguish the menu from—say a Delhi's restaurant. While this argument does not call for an arrested stage of development, discouraging locals from capitalizing upon the avenues opened by tourism but points toward the neocolonialism induced by the urban dwellers (who flaunt their imperialism through economic and cultural capital). It would be best if the rate of development should better be decided by its inhabitants so that they would not struggle with the Ship of Theseus paradox, while seeking authentic place, culture and self for themselves and for those who are lured by the place's "authenticity."

Hence these practices, if continued, can have a deteriorating effect on the ecological conditions in Uttarakhand. While some second homeowners do raise the concern over the issues cited above saying "*Pahaado ke liye aaye, pahaad hi kharab kr rahe h* (They came for the mountains and are now ruining those very mountains)", but their concerns take the front seat only when such practices change or degrade the geography to an extent that it in turns affect the place's aesthetic and feel. Such short-term and hedonistic concerns are like no concerns raised at all.

9.5 Conclusion

The mountains of Uttarakhand receiving huge investments from second home aspirants have been making a cry for help for the betterment of the aspirants, the locals and the geography. It is immensely important to not just flag the practices of *casual urbanism* rooted in the logic of consumption because "when land, or more broadly nature, is subject to commodification then it can no longer support the basic necessities for human life" (Burawoy 2015: 19). While pointing to current lifestyle and its potential to create future vulnerabilities, an effort has been made to present this otherwise phenomenon from multiple lenses. The delicate balance that sustains both the land and the dream woven around the mountains trembles under the weight of unregulated desire. As we delve deeper into this narrative, the logic is to shift the relation of exploitation or casualness to a more symbiotic relationship of coexistence. The tale of second homes in the mountains of Uttarakhand extends far beyond personal choices and idiosyncratic meaning making and does not underscore the interconnectedness of the phenomenon to other life forms. Hence, this is not mere exploration into lifestyles; it is a reverberation of warning from the mountains. The very terrain that nurtures aspirations has the power to shape vulnerabilities if not approached with sensitivity and foresight. While creating Uttarakhand as an ideal tourist hotspot for domestic and foreign tourists, formulation of regulations that safeguard the harmony of human dreams and ecological realities is of profound significance.

References

Adamaik C (2016) Cottage sprawl: spatial development of second homes in Bory Tucholskie. Poland 147:96–106

Adger W, Barnett J, Brown K (2013) Cultural dimensions of climate change impacts and adaptation. Nat Clim Chang 3(2):112–117

Adongo CA, Taale F, Adam I (2018) Tourists' values and empathic attitude towards sustainable development in tourism. Ecol Econ 150:251–263

Barke M (1988) Second homes in the Balearic Islands. Geography 73:143–145

Baudrillard J (2009) The precession of Simulacra. In: Storey J (ed) Cultural theory and popular culture: a reader, 4th edn. Pearson, Harlow, pp 409–415

Benson M, O'Reilly K (2009) Lifestyle migration: expectations, aspirations and experiences. Ashgate Publication, London
Bloch N (2020) Beyond a sedentary Other and a mobile tourist: transgressing mobility categories in the informal tourism sector in India. Crit Anthropol 40(2):1–20
Bourdieu P (1977) Outline of a theory of practice. Nice R (trans). Cambridge University Press, Australia
Burawoy M (2015) Facing an unequal world. Curr Sociol 63(1):5–34
Chakrabarty D (2012) Postcolonial studies and the challenge of climate change. New Literary History 43(1):1–18
Charles KK, Hurst E, Roussanov N (2009) Conspicuous consumption and race. Quart J Econ 124:425–467
Clifford J (1997) Routes. Travel and translation in the late twentieth century. Harvard University Press, London
Department of Economics and Social Affairs (2019) World urbanization prospects: The 2018 Revision. https://population.un.org/wup/Publications/Files/WUP2018-Report.pdf. Accessed 1 Aug 2023
Ericson T, Kjønstad BG, Barstad A (2014) Mindfulness and sustainability. Ecol Econ 104:73–79
Fiske SJ, Crate SA, Crumley CL, Galvin K (2014) Changing the atmosphere: anthropology and climate change. In: Final Report of the AAA Global Climate Task Force. American Anthropological Association, Arlington, VA
Gallent N (2007) Second homes, community and hierarchy of dwelling. Royal Geographical Society 39(1):97–106
Gallent N, Mace A, Tewdwr-Jones M (2005) Second homes: european perspective and UK policies. Ashgate, Aldershot
Ghertner DA, Lake RW (2021) Introduction. In: Ghertner DA, Lake RW (eds) Land fictions: The commodification of land in city and country. Cornell University Press, Ithaca, pp 1–25
Goodall B (1987) Dictionary of human geography. Penguin, Harmondsworth
Hall CM, Müller DK (2004) Tourism, mobility and second home: between elite landscape and common ground. Channel View Publication, Clevendon
Hannerz U (2002) Transnational connections: culture, people, places. Routledge, New York
Heffetz O (2011) A test of conspicuous consumption: visibility and income elasticity. Rev Econ Stat 93:1101–1117
Heffetz O (2012) Who sees what? Demographics and the visibility of consumer expenditures. J Econ Psychol 34(4):801–818
Hoggart K (1997) Rural migration and counter-urbanization in the European periphery: the case of Adalucia. Sociol Rural 37(1):134–153
Hoogendoorn G, Visser G (2004) Second homes and small-town (re)development: the case of Clarens. J Fam Ecol Consum Sci 32:105–115
Hoogendoorn G, Visser G (2010) The role of second homes in local economic development in five small South African towns. Dev South Africa 27(4):547–562
Kalternborn BP (1998) The alternate home: motives of recreation home use. Norsk Geografisk Tidskrift 52:121–134
Kirsch S (2001) Lost worlds: environmental disaster, "Culture Loss" and the law. Curr Anthropol 42(2):167–198
Makimoto T, Manners D (1997) Digital nomad. Wiley, New York
Mathur N (2016) Paper Tiger: law, bureaucracy and the developmental state in Himalayan India. Cambridge University Press, Delhi
Mi Z, Guan D, Liu Z, Liu J, Wang Y (2018) Cities: the core of climate change mitigation. J Clean Prod 207(1):258–285
Miller D (2013) Consumption and its consequences. Polity Press, Cambridge, UK
Moss L (2006) The amenity migrants-seeking and sustaining mountains and their culture. CABI, Wallingford, United Kingdom

Müller DK (1999) German second home owners in the swedish countryside: on the internationalization of the leisure space. European Tourism Research Institute, Sweden
Müller DK (2002) Second home ownership and sustainable development in northern Sweden. Tour Hosp Res 3(4):343–355
Nakano G, Rautela P, Shaw R (2017) Uttarakhand disaster and land use policy changes. In Banba M, Shaw R (eds) Land use managment in disaster risk reduction, pp 237–252
Nelson DR, Adger WN, Brown K (2007) Adaptation to environmental change: contributions of a resilience framework. Annu Rev Environ Resour 32:395–419
Osbaldiston N (2012) Seeking authenticity in place, culture and the self: the great urban escape. Palgrave Macmillan, New York
Osbaldiston N, Picken F, Duffy M (2014) Characteristics and future intentions of second homeowners: a case from Eastern Victoria, Australia. J Policy Res Tourism 7(1):62–76
Parkins W, Craig G (2006) Slow living. Berg, Oxford
Patterson T (1994) Toward a properly historical ecology. In: Crumley CL (ed) Historical ecology: cultural knowledge and changing landscapes. School of American Research Press, Santa Fe, NM
Pelling M (2011) Adaptation to climate change: from resilience to transformation. Routledge, London, UK
Pine B, Gilmore JH (1999) The experience economy. Harvard Business School Press, Boston, MA
Ragatz RL (1977) Vacation homes in rural areas: towards a model for predicting their distribution and occupancy patterns. In: Coppock JT (ed) Second homes: curse or blessings. Pergamon, Oxford, pp 181–193
Rudiak-Gould P (2014) Climate change and accusation: global warming and local blame in a small island state. Curr Anthropol 55(4):365–386
Saxena AK, Roy B (2022) New tourism paradigm In Uttarakhand tourism: innovations, dynamic and future perspectives. In: Khan AL, Hamid S, Sreedharan A (eds) New tourism paradigm in a changing world: innovations, dynamics, and future perspectives. Bharti Publication, New Delhi, pp 1–10
Schipper E (2020) Maladaptation: when adaptation to climate change goes very wrong. One Earth 3(4):409:414
Schonfeld B, Winter-Levy S (2020) Factual or moral persuasion in the United States? Evidence from the Papal encyclical on climate change. Anthropology of Religion ejournal. https://doi.org/10.2139/ssrn.3579152
Sheller M, Urry J (2006) The new mobility paradigm. Environ Plan 38:207–226
Singh K (2023) Buying property in Uttarakhand to cost more as circle rates soar. The Times of India. https://timesofindia.indiatimes.com/city/dehradun/buying-property-in-uttarakhand-to-cost-more-as-circle-rates-soar/articleshow/97995797.cms?from=mdr. Accessed 12 Sept 2023
Swim JK, Stern PC, Doherty TJ, Clayton S, Reser JP, Weber EU, Gifford R, Howard GS (2011) Psychology's contribution to understanding and addressing global climate change. Am Psychol 66(4):241–250
Tombaugh LW (1970) Factors influencing vacation home locations. J Leis Res 2(1):54–63
Tyagi T (2022) Dehradun: 'Builders misusing foothills policies to erect buildings'. The Times of India. https://timesofindia.indiatimes.com/city/dehradun/builders-misusing-foothill-policies-to-erect-buildings/articleshow/93916946.cms. Accessed 18 Aug 2023
Uttarakhand Tourism Development Board (2019) Tourism & Hospitality Sector Profile. Government of Uttarakhand. https://investuttarakhand.uk.gov.in/themes/backend/uploads/IP-UK-Tourism%20Sector%20Profile_V3_2019-11-15.pdf. Accessed 18 Aug 2023
Veblen T (1899) The theory of leisure class. Oxford University Press, Oxford
Vincentnathan L, Vincentnathan SG, Smith N (2016) Catholics and climate change skepticism. Worldviews 20(2):125–149
Visser G (2003) Visible, yet unknown: reflections on second home development in South Africa. Urban Forum 14(4):379–407

Wang J, Xia L (2021) Revenge travel: nostalgia and desire for leisure travel post COVID-19. J Travel Tour Market 38(9):935–955

Wisner B, Blaikie P, Cannon T, Davis I (2004) At risk: natural hazards, people's vulnerability, and disaster, 2nd edn. Routledge, London, UK

Wu Y, Xu H, Lew AA (2015) Consumption-led mobilized urbanism: socio-spatial separation in the second-home city of Sanya. Mobilities 10(1):136–154

Chapter 10
Exploring the Potentials of Community Participation in Landslide Risk Reduction: A Case Study of Dumsi Pakha in the District of Kalimpong, West Bengal

Phup Kesang Bhutia

Abstract The local communities play a crucial role in addressing the immediate needs, during and after a disaster. Community involvement and participation has thus become increasingly important in the disaster mitigation process. Despite the importance of community involvement in risk reduction, the biggest obstacle in any hazard-prone area is the absence of local readiness toward preparedness for disaster prevention and recovery. This chapter therefore investigates the impact or lack of community involvement and participation in the disaster mitigation process by taking the case study of Dumsi Pakha—a small compact village in the district of Kalimpong. Trapped between two sizeable natural drains (*jhoras*) and lacking any proper drainage infrastructure, Dumsi Pakha is vulnerable to significant erosion and slope instability resulting in landslides. A multiple regression model was used to analyze the interplay between socio-economic factors, disaster preparedness and mitigation measures like following updates regarding disaster-related information, participation in awareness programs, neighborhood relationship and community participation. The results show that while socio-economic factors are important, they share a weak correlation with disaster mitigation measures. Gender is a significant factor influencing community participation where females are notably more active in following landslide-related updates and participating in landslide-related awareness programs. In addition, there is a positive correlation between higher educational attainment and active participation in disaster mitigation measures indicating that education is a key driving force in enhancing disaster awareness and community involvement.

Keywords Community participation · Disaster mitigation · Landslide · Dumsi Pakha · Kalimpong

P. K. Bhutia (✉)
Department of Geography, Cluny Women's College, Kalimpong, India
e-mail: pkbclunycollege@gmail.com

A. Sarkar et al. (eds.), *Risk, Uncertainty and Maladaptation to Climate Change*,
Disaster Risk Reduction, https://doi.org/10.1007/978-981-99-9474-8_10

10.1 Introduction

The world is experiencing disasters at an unprecedented scale. Countries worldwide have witnessed a multitude of disasters along with their adverse effects and impacts on people (Behera 2021). Whether a disaster is categorized as a national disaster, major, minor or local disaster, it is the individuals within the communities who invariably bear the brunt of its adverse effects (Victoria 2003). In the year 2022, the Emergency Event Database (EM DAT) documented a total of 387 natural disasters and hazards globally (CRED Crunch 2023). Disaster Risk Reduction (DRR) policies, both at the national and state levels acknowledge the crucial role of communities as first responders during crises and underscore the significance of local expertise and adaptive strategies (Masson 2015). Community involvement and participation has thus become increasingly important in the disaster preparedness and mitigation process. With the rapid rise in disasters around the world, it becomes pertinent to examine the disaster management policies and methods with special emphasis on the role of communities as the most significant stakeholders during and after a disaster.

The World Health Organization (WHO) has defined Community participation as "the active involvement of people from communities preparing for or reacting to disasters. True participation entails the active engagement of the affected individuals in processes such as analysis, decision-making, planning and program execution. It also encompasses their involvement in all activities right from search and rescue to reconstruction, without the direct intervention of any external organizations" (Wisner et al. 2002). The concept of community-based disaster management gained prominence as an alternative approach during the 1980s and 1990s. In this approach, the local community is given primary attention in disaster reduction efforts as it represents the basic unit which is most commonly impacted by disasters and is significantly the first to respond and address the event (Dharmasena et al. 2008). Community participation is crucial for decreasing vulnerability to disasters, expediting recovery post-disaster and fostering community organization which serves as the foundation for sustainable development (Wisner et al. 2002).

Thus, the last decade has witnessed the advocacy of a paradigm shift in addressing and mitigating the adverse impact of disasters on human life and property (Victoria 2003). Contradictory to the reactive, top-down approach that primarily relies on structural and technological solutions, the new approach emphasizes the initiatives that engage local communities who often experience the most severe impacts of disasters. This modern approach places the responsibility for disaster mitigation on the communities themselves (Victoria 2003). The reduction of vulnerability is not solely dependent on the physical measures to mitigate the damaging effects of a hazard. Social measures aimed at diminishing adverse consequences and enhancing the population's resilience are equally vital. The effectiveness of any technical interventions, whether they occur before or after a disaster, hinges on how it is accepted and applied by the community. Therefore, community participation is a crucial component of emergency management planning (Wisner et al. 2002). Local communities have the opportunity for more substantial participation in shaping decisions that impact their lives, granting them a sense of control over their natural and physical surroundings (Victoria 2003). The major objective of Community Based Disaster Management (CBDM) is thus to transform and empower vulnerable or at-risk communities to disaster resilient ones (Victoria 2003). The incorporation of community disaster risk reduction into local development planning systems and processes will hence result in the creation of a sustainable and equitable community development (Victoria 2003).

Acknowledging that disasters stem from vulnerability influences strategies for building resilience and directs attention toward enhancing community empowerment and capacity development (Lawry and Cavalho 2014). Communities with a strong sense of solidarity, trust and active participation in the community tend to respond better to disasters (Choo and Yoon 2022). Disaster preparedness, prevention and recovery have been linked with higher levels of social participation including trust, positive and close-knit neighborhood relationships and active involvement in social organizations (Aldrich and Crook 2008; Reininger et al. 2013). The social bonds that develop between close neighbors facilitate collaborative and efficient decision-making contributing to the strengthening of social capital and swift response in the face of disaster (Choo and Yoon 2022).

This study is conducted using Dumsi Pakha—a small compact village in the district of Kalimpong, West Bengal as a case study and exploring the potentials of community participation and its influence on landslide risk reduction. The study analyzes the demographic and socio-economic factors that influence their ability to be resilient and reduce risks to landslide disasters (Figs. 10.1 and 10.2).

10.2 Methodology

The study attempts to explore the potential of community participation in Dumsi Pakha (Kalimpong District, West Bengal) and the impact it has on disaster risk reduction. For this, a multivariate regression model was used with a sample size of 145 households to examine the relationship between disaster mitigation measures such as awareness or participation in awareness events on landslides, landslide-related updates, relationship with neighbors and importance of community participation. A purposive sampling method was employed in Dumsi Pakha, to select households based on their increased vulnerability to landslides.

A questionnaire-based household-level survey was conducted. The questionnaire comprised of two sections. The first section focused on gathering data on the socio-economic characteristics of respondents like age, gender, caste, religion, educational level, employment status, household size, families with special needs members, quality of housing and access to media devices. The second half of the questionnaire centered on disaster mitigation variables like accessing landslide-related updates, participation in disaster (landslide) awareness programs, neighborhood relationship and community participation. The variables that have been taken are in view of the local set-up of the study area to reflect the conditions of vulnerability, particularly

Fig. 10.1 Site Photograph 1—Dumsi Pakha

Fig. 10.2 Site Photograph 2—Dumsi Pakha

social vulnerability. The concept of social vulnerability has gained widespread acceptance as a valuable tool for identifying individuals and places that are at risk and in need for disaster preparedness, response, recovery and mitigation measures (Park and Xu 2021). The variables mentioned above have thus been taken considering the basic domains of social vulnerability which are (1) socio-economic status, (2) household composition and disability, (3) minority status and (4) housing conditions (Flanagan et al. 2011). Multivariate regression analysis was used to analyze the data to derive at the conclusion.

The descriptive statistics (Table 10.1) reflects a rather balanced representation of gender with nearly equal number of male and female respondents. The mean age of population of 18 years or older for Dumsi Pakha was 40 years. The population of Dumsi Pakha was concentrated in the age group of 26–49 years.

The caste system is a social structure/hierarchy that has historically existed in South Asian societies including India and Nepal that classifies individuals into distinct social groups or castes based on factors like birth, occupation and social status. It is important to note that the caste system in India has undergone significant reforms both socially and legally and the terms SC, ST, OBC and General are used in official government classifications and policies to promote social equity and representation. The SC category encompasses the historically disadvantaged groups or the ex-untouchables, the ST category includes Indigenous or tribal communities (marginalized population) with their unique culture, language and traditions, the OBC category includes other socially and educationally backward classes and the general category includes individuals who do not belong to the SC, ST and OBC categories. In this contextual background, understanding these terms becomes crucial as they reflect specific social and demographic characteristics within the population in Dumsi Pakha. The caste composition in Dumsi Pakha showed diversity in terms of the distribution of the population with 32% belonging to the general category, 26.2% OBC, 22.8% SC and 18.6% ST.

Education was categorized into four levels ranging from those who could not read and write, those who could read and write and those who completed high school up to graduate level. Only 11% of the population had completed college while 24% were illiterate.

The households were also assessed based on the presence of special needs people in order to understand the challenges faced by them during situation of distress and disaster. The employment status was measured by simply classifying the population into employed and unemployed groups which provides a summary of the employment state. However, this approach is not useful in representing the complexity of the workforce as it fails to take into account the diverse range of employment situations. The household size was measured by categorizing it into single occupancy households, double occupancy households and households with 3/4/5 and more than 5 members. It is seen from Table 10.1 that 31.7% of the total population lived in households with more than 5 members. The households were also categorized into APL households and BPL households with 82.1% of the total households belonging to the BPL category. Lastly, the ownership of media devices showed 89.7% of the total population owning some form of media device like television, radio or phones.

Table 10.1 Descriptive Statistics of the respondents of the field survey

Variable name	Frequency	Percentage
Gender		
Male	71	49
Female	74	51
Caste composition		
Scheduled caste (SC)	33	22.8
Scheduled tribe (ST)	27	18.6
Other backward classes (OBC)	38	26.2
General	47	32.2
Age group		
18–25	15	8
26–33	40	21.3
34–41	36	19.1
42–49	20	10.6
50–57	9	4.8
58–65	14	7.4
65+	11	5.9
Educational qualification		
Who cannot read and write	35	24.1
Who can read and write	75	51.7
High school	19	13.1
Graduate	16	11
Special needs persons	32	22.1
Employment status		
Employed	67	48.2
Unemployed	78	53.8
Household size		
1	1	0.7
2	8	5.5
3	37	25.5
4	29	20.0
5	24	16.6
5+	46	31.7
Household category		
Above poverty level (APL)	26	17.9
Below poverty level (BPL)	119	82.1
Ownership of media devices	130	89.7

Source Author's calculation from primary survey

10.3 Results and Discussions

10.3.1 Study Area

Dumsi Pakha is a densely populated compact village in the district of Kalimpong, West Bengal, India, having an area of approximately 1.27 acres. It is a hilly area with high density housing and poor or no surface water management (Anderson et al. 2013). Kalimpong is characterized by steep slopes, loose topsoil, undulating terrain and hillsides with an average elevation of 2050 m and experiencing heavy rainfall rendering it extremely vulnerable to landslides (Roy et al. 2022; Anderson et al. 2013).

The susceptibility of a population or area to hazards is influenced by both its geographical location and social conditions (Logan and Xu 2015). Therefore, it is very important to understand the socio-economic dynamics of a region or place in order to learn and explore its potential in terms of disaster risk reduction and management.

Falling under the purview of Kalimpong Municipality, it provides a unique case study representing both rural and urban vulnerability to landslides. The population of Dumsi Pakha consists of individuals from diverse socio-economic backgrounds with a majority of population belonging to the general category. The gender distribution in Dumsi Pakha is fairly balanced while the age distribution reflects a concentration of population between the age group of 26–49 years constituting 51% of the total population. The levels of educational attainment indicate a relatively low proportion of individuals with college degrees while approximately 27% of the population remains illiterate signifying a considerable percentage of adults who lack basic reading and writing skills. A majority of the population falling into the BPL category (more than 80%) implies economic disparities among the population and the prevalence of economic challenges or low-income status (Fig. 10.3).

All the disaster mitigation measures namely neighborhood relationship, following landslide-related updates, participation in landslide-related awareness programs, significance of community's role in disaster risk reduction and community participation were subjected to a multivariate regression analysis that took into account relevant socio-economic variables.

When disaster strikes, it is the local community that experiences it first and responds promptly, even before external help arrives (Choo and Yoon 2022). From Table 10.2, it is clearly seen that caste has a significant positive correlation with neighborhood relationships. This indicates that caste individually contributes significantly to the prediction of neighborhood relationship. This further implies that people belonging to the general categories are likely to share extremely good relationships with their neighbors while people belonging to the reserved categories, notably the Scheduled Caste (SC) category, are less likely to mingle and have poor relationships with their neighbors.

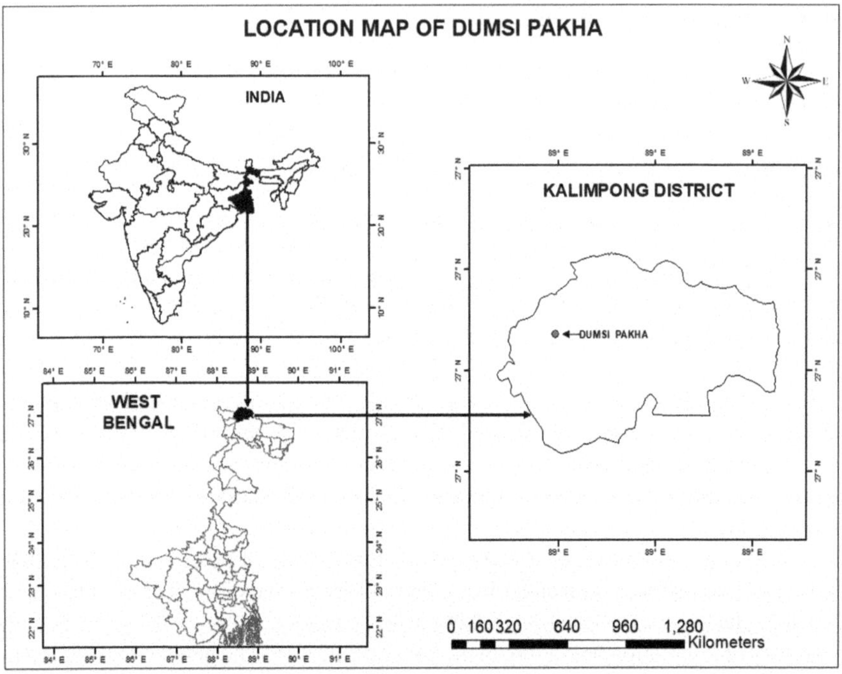

Fig. 10.3 Location map showing Dumsi Pakha

Table 10.2 Community participation and neighborhood relationship

Predictor	Beta coefficient	t-value
Gender	0.036	0.358
Age	0.116	1.240
Caste	0.203	2.123*
Educational qualification	0.031	0.331
Employment status	−0.038	−0.370
Any special needs person in the family?	−0.002	−0.025
Household size	−0.100	−1.120
APL or BPL category	−0.105	−1.187

*Statistically significant at 95% confidence level ($p < 0.05$)
Source Author's calculation from primary survey

Table 10.3 shows the significance of gender, educational qualification and employment status in determining whether individuals closely follow landslide-related updates. Females tend to be more active in following landslide-related updates than their male counterparts. Positive and significant coefficient associated with educational qualifications suggest that individuals with higher education levels are more active in following landslide-related updates and information. A negative and significant coefficient for employment status indicates that people who are unemployed are less likely to follow any landslide-related updates.

Table 10.4 shows a negative and significant relationship of families with special needs persons with the importance of community in disaster risk reduction. This indicates that the presence of special needs persons as a variable contributes significantly in the prediction of the importance of the role of community in disaster risk reduction. It demonstrates that the role of community in disaster risk reduction becomes increasingly important particularly among disabled persons.

From Table 10.5, it can be seen that gender and educational qualification are positive and significant indicating that these variables individually contribute significantly in the prediction of participation of individuals in disaster/landslide-related awareness programs. A positive and significant coefficient of educational level reflects that

Table 10.3 Community participation and population following any disaster-related updates

Predictor	Beta coefficient	t-value
Gender	0.183	2.015*
Age	−0.067	−0.790
Caste	−0.022	−0.254
Educational qualification	0.292	3.462*
Employment status	−0.357	−3.838*
Any special needs person in the family?	−0.126	−1.372
Household size	−0.006	−0.078
APL or BPL category	−0.083	−1.038

*Statistically significant at 95% confidence level ($p < 0.05$)
Source Author's calculation from primary survey

Table 10.4 Community participation and significance of community's role in disaster risk reduction

Predictor	Beta coefficient	t-value
Gender	0.060	0.603
Age	0.027	0.285
Caste	0.024	0.259
Educational qualification	0.051	0.547
Employment status	−0.060	−0.578
Any special needs person in the family?	−0.237	−2.405 *
Household size	0.034	0.391

*Statistically significant at 95% confidence level ($p < 0.05$)
Source Author's calculation from primary survey

Table 10.5 Community participation and involvement in disaster awareness programs

Predictor	Beta coefficient	t-value
Gender	0.210	2.135*
Age	−0.114	2.135
Caste	−0.080	2.135
Educational qualification	0.265	2.135*
Employment status	−0.184	2.135
Any special needs person in the family?	0.075	2.135
Household size	−0.070	2.135

*Statistically significant at 95% confidence level ($p < 0.05$)
Source Author's calculation from primary survey

participation in disaster awareness programs increases with higher educational qualification. Similarly, gender exhibits a positive and significant coefficient explaining that females are more active in participation in disaster/landslide-related awareness programs.

10.3.2 Findings

The present study investigates and analyzes the relationship between socio-economic characteristics and the different disaster mitigation measures of Dumsi Pakha—a village that has been rightly described by Save the Hills as living on borrowed time (Rao 2010).

The major finding of this study is that the socio-economic parameters taken to analyze and explore the potentials of awareness and participation in disaster/landslide risk reduction in Dumsi Pakha, Kalimpong are not highly correlated. It is clearly seen from the empirical analysis that out of the different variables like age, gender, caste, educational qualification, employment status, presence of special needs persons, household size and whether APL or BPL cardholders were used to understand and examine neighborhood relationships; it was only caste differences that influenced neighborhood relationships. Persons belonging to the SC categories lacked social capital and hence were more vulnerable to landslides. This finding reflects the challenges associated with the caste-based dynamics that seem to be strongly rooted among the population of Dumsi Pakha, Kalimpong. These dynamics also serve as a significant factor in determining vulnerability, creating distinctions between who is most vulnerable and who is not (Bosher et al. 2007). However, it is important to note that the influence of caste on neighborhood relationships is likely to be different across regions and thus the result from this study is context-specific. Here, caste appears to play a pivotal role in influencing social interactions and relationships with the neighbors that may adversely affect community participation and resilience to disasters. The result thus emphasizes the need for social awareness, community-building programs and formulation of policies that would promote inclusiveness among communities which would in turn benefit successful disaster risk reduction and mitigation measures because good neighborhood relationships are critical during times of distress (Ali and George 2021).

The educational status and gender differences contributed in influencing the individual's interest on disaster-related updates and participation in landslide awareness programs. The observed gender differences indicate women to be more involved in landslide-related updates with a higher rate of female participation in landslide-related awareness programs. This increasing involvement of females in the various disaster mitigation measures indicates a possible gender difference in hazard awareness and information-seeking behavior. This finding opens up newer avenues for further research on gender dynamics in terms of their response to disaster preparedness and recovery. Women have a significant role to play during disaster relief and reconstruction activities (Gokhale 2008). Women's groups involved in emergency relief, resettlement and reconstruction efforts following a natural disaster acquire valuable knowledge and expertise that can greatly benefit communities experiencing similar crises in future (Gokhale 2008). Thus, gender can play an important role in reducing vulnerability and women play a greater role in these processes, largely operating at local levels (Bosher et al. 2007). Education and raising awareness are identified as key factors influencing the effectiveness of landslide disaster risk reduction (Klimes et al. 2019). A positive correlation between higher education levels and the disaster mitigation measures like following landslide-related updates and participation in landslide awareness programs suggest that education is vital in increasing disaster awareness and engagement among communities. Individuals with higher levels of education are more likely to access and respond to hazard information from the initial phases of preparedness right up to the recovery phase (Tierny 2006).

In the case of people with special needs and importance of the role of community in disaster risk reduction, it was evident that households with special needs persons played an important role in determining the importance of the role of community in disaster risk reduction. Many older or disabled people have special needs that require the assistance of others (Flanagan et al. 2011). Hence, this could be a major reason for the increasing importance of the role of community among people with special needs.

Despite Dumsi Pakha being a very vulnerable landslide-prone and affected region, it is evident that the role of community in landslide risk reduction is almost an unexplored subject. Moreover, the socio-economic indicators that are key aspects in predicting the nature and extent to which communities take measures to reduce the impact of disasters reflect a weak correlation. In the context of Dumsi Pakha, it becomes crucial to emphasize on the significance of social capital as one of the vital resources as it strengthens the community's ability to prepare for, respond to and recover from disasters (Dynes 2002; Aldrich 2012). This may therefore imply that along with positive socio-economic changes, focus should also be concentrated on raising awareness, providing accessible information, encouraging community engagement and delving deeper into the local context of the region and the associated disasters.

10.4 Limitations

The study takes into consideration only the selected socio-economic variables which may not have fully explained the variation in disaster mitigation measures. Moreover, the results are context-specific and cannot be generalized. This study did not include several factors typically regarded as contributors to social vulnerability such as access to resources (including information and political power), perception and awareness of risk and variables of community participation like trust, strong and weak ties of communities in access to resources and support during times of distress (Aksha et al. 2018; Ali et al. 2022). Furthermore, Dumsi Pakha being a small compact village, provided a relatively small (145) sample size limiting the statistical ability to identify significant relationships.

10.5 Conclusion

The study conducted in Dumsi Pakha, Kalimpong, is a unique case study as it reflects both the rural and urban vulnerability to landslides. This study finds that the socio-economic variables and disaster mitigation measures share a weak correlation indicating the need for a more nuanced approach to disaster preparedness and mitigation. Such weak linkages suggest that further research is required to understand the underlying components that contribute to community participation in disaster preparedness and mitigation.

References

Aksha SK, Juran L, Lynn MR, Zhang Y (2018) A analysis to social vulnerability to natural hazards in Nepal using a modified social vulnerability index. Int J Disaster Risk Sci 10:103–116. https://doi.org/10.1007/s13753-018-0192-7

Aldrich DP, Crook K (2008) Strong civil society as a double edged sword: sitting trailers in post-Katrina New Orleans. Polit Res Q 61(3):379–389. https://doi.org/10.1177/1065912907312983

Aldrich DP (2012) Building resilience: social capital in post-disaster recovery. The University of Chicago Press, Chicago

Ali S, George A (2021) Social inclusivity: a case study on community resilience on Kerala Flood-2018. Lecture notes in civil engineering. Springer, Singapore, pp 109–131

Ali S, George A (2022) Fostering disaster mitigation through disaster mitigation-case of Kochi residents following the Kerala floods of 2018 and 2019. Springer Nat 111(1):389–410. https://doi.org/10.1007/s11069-021-05058-0

Anderson MG, Elizabeth H (2013) Community-based landslide risk reduction: managing disasters in small steps. The World Bank, Washington DC

Behera JK (2021) Role of social capital in disaster risk management: a theoretical perspective in special reference to Odisha, India. Int J Environ Sci Technol 20:3385–3394. https://doi.org/10.1007/s13762-021-03735-y

Bosher LS, Penning-Rowsell E, Tapsell S (2007) Resource accessibility and vulnerability in Andhra Pradesh: caste and non-caste influences. Dev Chang 38(4):615–640

CRED Crunch (2023) Disasters year in review 2022. Issue no 70 http://cred.be/sites/default/files/2022EMDATreport.pdf

Cutter SL, Boruff BJ, Shirley WL (2003) Social vulnerability to environmental hazards. Soc Sci Q 84(2):242–261

Dharmasena PB, Jayathilaka KMDP (2008) Community based disaster risk reduction modeling. Sri Lanka Foundation Institute, Oxfam America, Institute for Human Development and Training

Dynes RR (2002) The importance of social capital in disaster response. University of Delaware, Disaster Research Centre

Dynes RR (2006) Social capital dealing with community emergencies. Homeland Secur Affairs 2(2):1–26. http://hdl.handle.net/10945

Flanagan BE, Gregory EW, Hallisey EJ, Heitgerd JL, Brian L (2011) A social vulnerability index for disaster management. J Homeland Secur Emerg Manage 8(1–3). https://doi.org/10.2202/1547-7355.1792

Gokhale V (2008) Role of women in disaster management: an analytical study with reference to Indian society. In: The 14th world conference on earthquake engineering, Beijing, China

Kelman I, Jessica M, Gaillare JC (2012) Indigenous knowledge and disaster risk reduction. Geograph Assoc 97(1):12–21

Klimes J, Calvello M, Auflic MJ (2019) Objectives and main results of "community participation for landslide disaster risk reduction' thematic papers. Springer-Verlag, GmBH Germany, Doi. https://doi.org/10.1007/s10346-019-01246-z

Lawry JB, Cavalho L (2014) Building local level engagement in disaster risk reduction: a Portuguese case study. Emerald Group Publishing. https://doi.org/10.1108/DPM-07-2014-0129

Logan JR, Xu Z (2015) Vulnerability to hurricane damage on the U.S. gulf coast since 1950. Geogr Rev 105:133–155. https://doi.org/10.1111/j.1931-0846.2014.12064.x

Masson VL (2015) Considering vulnerability in disaster risk reduction plans: from policy to practice in Ladakh, India. Mount Res Dev 35(2):104–114

Mijn C, Yoon DK (2022) Examining the effects of the local communities' social capital on disaster response capacity in Seoul, South Korea. Int J Disast Risk Reduct 75. https://doi.org/10.1016/j.ijdrr.2022.102973

Park G, Xu Z (2021) The constituent components and local indicator variables of social vulnerability index. Springer Nature. https://doi.org/10.1007/s11069-021-04938-9

Rao P (2010) Dumsi Pakha (Kalimpong)- a village living on borrowed time. Save The Hills

Reininger BM et al (2013) Intention to comply with mandatory hurricane evacuation orders among Hispanics living along a Coastal area. Disast Med Public Health Prep 7(1):46–54. https://doi.org/10.1001/dmp.2012.57

Roy P, Ghosal K, Paul PK (2022) Landslide susceptibility mapping of Kalimpong in Eastern Himalayan region using a Rprop ANN approach. J Earth Syst Sci 131:130. https://doi.org/10.1007/s12040-022-01877-2

Tierny K (2006) Social inequality, hazards and disasters. Risk and disaster: lessons from Hurricane Katrina, pp 109–128

Victoria LP (2003) Community-based disaster management in the Philippines: making a difference in people's lives. Philipp Sociol Rev 51:65–80

Wisner B, J Adams (2002) The nature of emergencies and disasters. In: Wisner B, Adams J (eds) Environmental health in emergencies and disasters: a practical guide. World Health Organization, pp 9–22

Chapter 11
Livelihoods of Farmers Vulnerable to Climate Change: Evidence from Drought-Prone Regions of India

Surendra Singh Jatav, Nathoo Bharati, and Pooja Rathore

Abstract Indian farmers are vulnerable to changing climate with unpredictable rainfall distribution, rising temperature, and complex socioeconomic conditions. The present study aims to assess livelihood vulnerability of farmers in two regions of the most populous State of Uttar Pradesh namely Bundelkhand and Central region. By using multistage random sampling technique, a total of 480 samples from 16 villages, 8 development Blocks, 4 districts, and 2 regions were selected to elicit grass-root information on farmers' perception of climate change, their sensitivity, and adaptive capacity to changing climate. Further, this study has adopted four methodologies from the Intergovernmental Panel on Climate Change's Fourth Assessment Report for the development of climate vulnerability indices. The results show that farmers in Jhansi district were highly vulnerable to changing climate, while farmers in Barabanki district were relatively less vulnerable. The elevated degree of vulnerability to livelihood in Jhansi district attributed to its comparatively higher exposure and sensitivity to climatic change. Hence, the present study suggested that farmers, specifically in developing countries like India, must adapt to climate change to reduce its negative impact and reap the benefits of adaptation. This could be achieved through establishing training activities, skills development, and capacity to strengthen the farmers' ability to adapt. These programs are important not only to farmers but also for government officials to provide appropriate technical support to farmers.

Keywords Adaptive capacity · Climate change · Farmers perception · Indicator approach · Livelihood vulnerability index · Rainfed region · Sensitivity

---

S. S. Jatav (✉) · N. Bharati
Department of Economics, Babasaheb Bhimrao Ambedkar University, Lucknow, India
e-mail: surendra.singh735@gmail.com

P. Rathore
Guru Gobind Singh Indraprastha University, New Delhi, India
e-mail: poojarathore@ipu.ac.in

A. Sarkar et al. (eds.), *Risk, Uncertainty and Maladaptation to Climate Change*, Disaster Risk Reduction, https://doi.org/10.1007/978-981-99-9474-8_11

11.1 Introduction

Changes in the global climate, which have caught the attention of experts in every discipline, have far-reaching consequences for human civilization (Singh 2020a). The effect is varied in various places of the world (IPCC 2018). Negative effects are predicted to be more severe in tropical and subtropical climates, especially in developing nations like India (Singh et al. 2019). The average global temperature in 2022 was about 1.5 °C above the average temperature in 1850–1900 (World Meteorological Organization 2023). Further, the annual mean global near-surface temperature for each year between 2023 and 2027 is predicted to be between 1.0 °C and 1.8 °C higher than 1850–1900, with human activity being the primary driver of the observed warming as observed by the World Meteorological Organization. According to the Intergovernmental Panel on Climate Change (2014), between 20 and 40% of the world's population has already experienced warming of more than 1.5 °C on a regional basis. This has significant impact on agricultural productivity. Hence, agriculture has emerged as a central topic of discussion within the discourse around climate change.

Consequently, the implications of climate change and the associated risks in developing countries like India are of a greater magnitude. Furthermore, it has been shown by Singh and Sanatan (2014) and Singh et al. (2019) that a significant proportion of Indian farmers (i.e., 85%) have little financial resilience to deal with changing climate.

In addition, there exists a socioeconomic disparity in the vulnerability of farmers to fluctuations in monsoon patterns (Singh 2020a). Farmers with low-quality soil or unfavorable watershed positions who can't afford or don't have access to irrigation are more at risk for crop failures brought on by dry spells (Kumar et al. 2018). Reductions in agricultural productivity have been linked to dry periods that occur during critical phases of plant development (Singh et al. 2019). When it comes to minimizing production, gaps brought on by dry periods in India, Sikka et al. (2018) stress the need of securing irrigation and keeping soil moisture. Despite significant reductions in greenhouse gas emissions as a component of a climate mitigation plan, it is anticipated that adverse impacts of climate change would persist and intensify in the next decades; hence, necessitating the urgent implementation of adaptation measures (Singh and Sanatan 2018a, b). Objectives, scope, and the degree to which the findings of climate vulnerability assessments may be put into practice can vary greatly among studies (Jatav 2020; Sanatan and Singh 2020; Kuchimanchi et al. 2021). The evaluations might be done to monitor changes in vulnerability or to identify existing or prospective hotspots as entry sites for climate adaptation action (Singh 2020a). Mapping may be done on a national scale or on a scale as small as a single town.

Circular rather than linear relationships exist between vulnerability, adaptive capability, and adaptation (Mertz et al. 2009). It has been suggested that vulnerability and adaptability are linked, with the former depending on the latter (Kelly and Adger 2000). People are less likely to be impoverished and be hungry, if they have access to stable income and a varied diet, for instance. This, in turn, increases their capability to react to pressures by shifting resource allocations, leaving behind, or migrating agricultural regions, depending on the severity of the threats they face. Therefore, individuals' capacity to effectively cope with potential future stressors is compromised as a result of previous injury. For instance, in the event that individuals depend on credit schemes to acquire drought-resistant crops and crop varieties as a component of the essential adaptive measures to mitigate the impact of drought, a complete loss of crops would not only lead to food insecurity but will also burden them with insurmountable debts. Hence, the implementation of credit schemes and the cultivation of novel crops may need the provision of "weather insurance," as has been experimentally assessed in many developing countries (Barnett and Mahul 2007). Moreover, Kelly and Adger (2000) asserted that the evaluation of individuals' genuine vulnerability is unattainable until adaptation has been implemented. Farmers have a crucial role in addressing climate change and possess a unique advantage in implementing essential adaptation strategies to mitigate the effect of climate change on their agricultural systems.

The recognition of climate change has been widely acknowledged as a fundamental need for implementing any kind of adaptation measures. According to Alam et al. (2017), farmers who possess knowledge on the actualities of climate change and its significant ramifications are more likely to express support for government measures (Alam et al. 2017). In order to evaluate the vulnerability of farmers to climate change and their subsequent adaptation strategies, it is important to first examine their perception on this issue. The challenge of decision-making under constraints is exacerbated for farmers due to the time lag between information gathering and implementation. Several scholars, including Madison (2007), Bryan et al. (2009), Nhemachena and Hassan (2008), and Singh (2020a), have highlighted the importance of farmers' perceptions toward climate change within the framework of farm-level adaptation. The climatic components include several phenomena, such as the occurrence of localized climate extremes, including floods and droughts and fluctuations in average temperatures and precipitation throughout different geographical areas. Tripathi and Mishra (2017) explored the views of climate change among farmers that are influenced by the prevailing weather conditions. Consequently, it is important to conduct research and ascertain the specific climatic factors that farmers consider in their assessments. Due to climate uncertainty, farmers place more weight on recent climatic occurrences as information (Jatav 2020). Short-term observations reveal increasingly chaotic and unpredictable interannual temperatures and climatic extremes. Historically, farmers have been worried with seasonal climate projections because of the short reaction time and narrow decision window afforded by such shifts. The previous research also claims that farmers' perceptions of climatic events may not ensure adaptation strategies (Bryan et al. 2009), since a number of variables may compromise their adaptability. Farmers' cognitive abilities differ from one another and are affected by factors such as age, education, gender, and geography (Deressa et al. 2009; Funk et al. 2019; Jatav 2020).

The scope of this work has been broadened to include supplemental irrigation planning on a watershed scale. Therefore, it is necessary to be able to identify and map relative vulnerability. Investments in life-saving irrigation access may be directed toward farmers who are more at risk of experiencing crop losses due to dry spells. This is true not just for India, but also for the worldwide phenomenon of more unpredictable rainy seasons in dryland rainfed areas. There is not, however, a method that can be used to determine how dry season vulnerability may be conceptualized and computed at the level of a farm, at the same time, mapped for all farmers at an aggregate scale such as a village in order to facilitate the implementation of policies. This research gap is addressed in this study.

The objective of this study is to analyze perception of farmers on climate change and to determine the factors that contribute to livelihood vulnerability in the Bundelkhand and Central regions of Uttar Pradesh, which is most populous State in India.

11.2 Methods and Materials

11.2.1 Sampling Technique and Sample Size

To collect field survey data, the current study employs a multistage random sampling procedure. Field survey was conducted in the months of August and September 2022. In the first step, Uttar Pradesh was purposely selected. In the second step, two regions were purposely selected from Uttar Pradesh to identify the drivers of livelihood vulnerability—first an under-developed region, Bundelkhand and second a developed region, Central. In the third step, two districts from each region were selected. The Jhansi and Lalitpur districts from Bundelkhand region, and Lucknow and Barabanki districts from Central region were selected. In the fourth step, two Development Blocks from each district were selected. In the fifth step, two villages (micro administrative unit) from each Development Block were selected. Lastly, 30 samples from each village were selected. Thus, 2 regions, 4 districts, 8 Development Blocks, 16 villages, and 480 samples were selected to capture farmers' perception of climate change and to identify the drivers of livelihood vulnerability in most populous State of India namely Uttar Pradesh (Fig. 11.1).

11.2.2 Estimation Method: Indicator Approach

The present study employs an indicator-based methodology to compute a livelihood vulnerability index for the farmers included in the sample. The indicator approach is widely employed and has numerous advantages that have facilitated its extensive use in the planning process and policy communication over time. One key advantage is its ability to condense a substantial amount of intricate information into a manageable format (Jatav 2022). Additionally, this approach allows for the utilization of data at

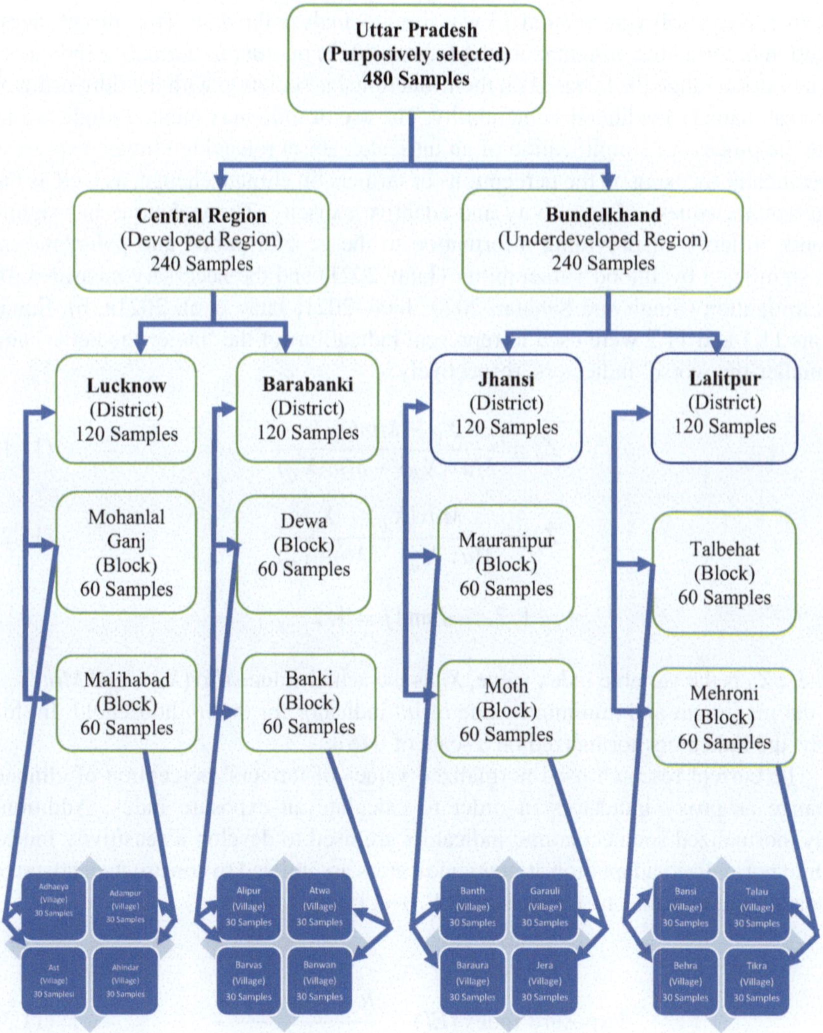

Fig. 11.1 Tree diagram of sampling technique. *Source* Authors' creation

various levels, ranging from individual to national, in order to construct a livelihood vulnerability index. Moreover, in situation where original data is unavailable, proxy data can be used. This approach also enables the identification, prioritization, and ranking of the climate vulnerable districts, thereby aiding in the identification of potential barriers to district development. Lastly, the indicator approach is valuable for monitoring and evaluating the effectiveness of interventions (Jatav 2020).

The present study used differential data to compute the livelihood vulnerability index, thereby necessitating the consideration of the normalization procedure.

Further, Excel software version 13 was used to analyze the data. The current investigation has used the min–max method (Jatav 2022) in order to normalize indicators to a uniform range (0, 1) based on their functional association with the dimension of interest, namely livelihood vulnerability. The use of min–max method might facilitate the process of simplification of an intricate dataset related to climate exposure, specifically focusing on the perceptions of farmers on climate change, as well as the interconnectedness of sensitivity and adaptive capacity. The technique has significance in terms of providing information to the general public and policymakers on significant livelihood vulnerability (Jatav 2021) and the necessary measures for its mitigation (Singh and Sanatan 2020; Jatav 2021; Jatav et al. 2021a, b). Equations 11.1 and 11.2 were used to represent indications of the 'larger-the better' and 'smaller-the-worse' indicators, respectively.

$$Z_{ij} = \frac{X_{ij} - Min(X_{ij})}{Max(X_{ij}) - Min(X_{ij})} \tag{11.1}$$

$$Z_{ij} = \frac{Max(X_{ij}) - X_{ij}}{Max(X_{ij}) - Min(X_{ij})} \tag{11.2}$$

$$i = 1, 2, \ldots I \text{ and } j = 1, 2, \ldots.$$

where Z_{ij} is the variable index value, X_{ij} is the actual value, $Max(X_{ij})$, and $Min(X_{ij})$ is the maximum and minimum value of *ith* indicator for the *jth* household. In this way, the indicators normalized on a scale of 0 to 1.

The current research used normalized values of farmers' perception of climate change as proxy indicators in order to calculate an exposure index. Additionally, normalized socioeconomic indicators are used to develop a sensitivity index, while normalized adaptation strategy indicators are utilized to construct an adaptive capacity index. Table 11.1 uses Eqs. (11.3)–(11.5) as follows.

$$\text{Exposure Index } (EI) = \frac{R + S + D + W}{4} \tag{11.3}$$

$$\text{Sensitivity Index } (SI) = \frac{F + DW + Ir + FHH + M + T + H + BPL + E + I}{10} \tag{11.4}$$

$$\text{Adaptive Capacity Index } (SI) = \frac{CPR + NF + JF + KKC + CWB + In + St + Tr + NPK + ASD + CD}{11} \tag{11.5}$$

where the variables on the right-hand side are the normalized version of the indicators listed in Table 11.1. Once the values for exposure, sensitivity, and adaptive capacity for the household level had been calculated, two contributing factors (exposure and sensitivity) were combined using Eq. 11.6 to obtain the household-level potential livelihood vulnerability index (Jatav 2020).

$$PLVI_h = Exposure\ Index_h - Sensitivity\ Index_h \tag{11.6}$$

where $PLVI_h$ is the potential livelihood vulnerability index score for the household h; $ExposureIndex_h$ is the calculated exposure score for the household h; and $SensitivityIndex_h$ is the sensitivity score for the household h. Further, adaptive capacity, represented by A_h in Eq. 11.7 was taken into consideration for the development of livelihood vulnerability index as follows:

$$LVI_h = \left(Exposure\ Index_h - Adaptive\ Capacity\ Index_h\right) * Sensitivity\ Index_h \tag{11.7}$$

PLVI and LVI were scaled so that -1 denotes the least vulnerable and 1 is the most vulnerable.

11.2.3 Selection of Rational Indicators for the Development of Livelihood Vulnerability Index

The Intergovernmental Panel on Climate Change (IPCC) defines vulnerability, as stated in its Fourth Assessment Report, as the probability of being susceptible to adverse conditions, as well as the challenges associated with adaptation and mitigation of the sudden risks, shocks, and undesirable events resulting from climate change and climate-induced hazards (IPCC 2014). The adverse impacts of extreme events are closely associated with climate variability and gradual changes in key climatic factors such as mean temperature, rainfall, and climate-related hazards such as cyclones, storms surges, sea level rising, flooding, and coastal vulnerability. These impacts can be understood through three inter-connected dimensions: (i) adaptive capacity (the ability to cope with sudden risk), (ii) sensitivity (the extent to which a particular area or population is affected by an extreme event), and (iii) exposure (the intensity of climatic variability and the factors that contribute to vulnerability) (Hahn et al. 2009; Hoque et al. 2019; Jatav 2020).

Previous research has used farmers' perception of climate change as a means to approximate an exposure index (Masud et al. 2017; Shreshtha et al. 2017; Funk et al. 2019). According to Masud et al. (2017) in Malaysia and Jatav (2020) in Bundelkhand region, a significant majority of farmers, over 90% have reported perceiving changes in the climate. Specifically, they have seen a notable rise in temperatures and a decrease in predictability of rainfall patterns. According to Shreshta et al. (2017), farmers possess a comprehensive understanding of detrimental effects of climate change on agriculture and their overall livelihoods. However, their ability to adjust

Table 11.1 Rational indicators for livelihood vulnerability index

Component	Indicators	Functional relationship with livelihood vulnerability index	Source
Exposure	HHs perceived that rainfall declined (Rainfall)	Positive	Masud et al. (2017), Jatav (2020)
	HHs perceived that summer days become hotter (Summer Days)	Positive	Shrestha et al. (2017), Jatav (2020)
	HHs perceive those frequencies of drought has increased (Drought)	Positive	Funk et al. (2019), Jatav (2020)
	HHs perceive that water level has declined (Water Level)	Positive	Omerkhil et al. (2020), Jatav (2020)
Sensitivity	HHs using only forest-based energy resources for cooking purposes (Forest)	Positive	Omerkhil et al. (2020), Jatav (2020)
	HHs using hand-pump (untreated) water for drinking (Drinking Water)	Positive	Miranda et al. (2011), Jatav (2020)
	HHs depends on government sources for irrigation (Irrigation)	Positive	Miranda et al. (2011); Jatav (2020)
	Female-headed households (FHH)	Positive	Abid et al, (2015), Jatav (2020)
	HHs using 108 free medical facilities (Medical)	Negative	Alam et al. (2017), Jatav (2020)
	HHs do not have toilet facilities (Toilet)	Positive	Alam et al. (2017), Jatav (2020)
	HHs do not have all seasonal house (House)	Positive	Alam et al. (2017), Jatav (2020)
	HHs belong to Below Poverty Line category (BPL)	Positive	Rai et al. (2008), Jatav (2020)
	HHs do not have electricity connection (Electricity)	Positive	Alam et al. (2017), Jatav (2020)

(continued)

Table 11.1 (continued)

Component	Indicators	Functional relationship with livelihood vulnerability index	Source
	Head of Household does not attained school (Illiterates)	Positive	Nadeem et al. (2009), Jatav (2020)
Adaptive capacity	HHs changes their cropping pattern (Cropping Pattern Change)	Negative	Masud et al. (2017), Jatav (2020)
	HHs switch to non-farm activities (Non- farm)	Negative	Masud et al. (2017), Jatav (2020)
	HHs live in joint family (Joint Family)	Negative	Masud et al. (2017), Jatav (2020)
	HHs using Kisan Call Centre for agro-advisory (KKC)	Negative	Masud et al. (2017), Jatav (2020)
	HHs started conservation of water bodies and soil to combat climate variability (CWB)	Negative	Masud et al. (2017), Jatav (2020)
	HHs secure their crop through crop insurance (Insurance)	Negative	Masud et al. (2017), Jatav (2020)
	HHs have storage capacity to procure agriculture products (Storage)	Negative	Masud et al. (2017), Jatav (2020)
	HHs have taken professional training on climate change combating (Training)	Negative	Masud et al. (2017), Jatav (2020)
	HHs aware about nitrogen, phosphorus, and potassium ratio (NPK)	Negative	Masud et al. (2017), Jatav (2020)
	HHs growing more than one cropping (Crop diversification)	Negative	World Bank (1997), Jatav (2020)

Source Adopted from Jatav (2020) and Field Survey Data, 2022 to these challenges is hindered by many factors such as limited access to advanced technologies, small land holdings, and persistent drought conditions. The current research utilizes the perception of farmers to construct an exposure index for sampled districts in the Bundelkhand and Central regions of Uttar Pradesh.

Sensitivity refers to the extent to which a system is influenced, either negatively or beneficially, by climate-related stimuli (IPCC 2001) as well as the capacity of a system to adopt to climate change from socioeconomic and ecological perspectives (IPCC 2007). The development of a sensitivity index for various districts in the Bundelkhand and Central regions was based on the use of socioeconomic data outlined in Table 11.1.

While the system may exhibit a notable degree of susceptibility or responsiveness to environmental stress and shocks, it would be inaccurate to categorize it as susceptible (Fellmann 2012). The vulnerability of a system is influenced by its adaptive capacity, which in turn adjusts to both exposure and sensitivity (Jatav 2020). The achievement and effectiveness of adaptation processes are influenced by the three crucial factors: (i) the prompt recognition and comprehension of climate variations and the corresponding requirement to implement adaptive measures; (ii) the presence of incentives and the capacity to adapt; and (iii) the necessity to modify agricultural practices in order to optimize benefits in response to changing climate conditions (Deressa et al. 2009). Additionally, Masud et al. (2017) proposed that the implementation of climate change adaptation measures is essential in order to mitigate its adverse effects and capitalize on the advantages of adaptation. The authors further proposed that a greater level of adaption may be attained by implementing training programs, fostering skill development, and enhancing the capability of farmers to adjust. Specialized training sessions have significant importance not just for farmers but also for government officials responsible for delivering suitable technical assistance to the farmers. Therefore, the adaptive capacity index was constructed using the extension education data provided in Table 11.1.

11.3 Results and Discussion

11.3.1 Farmers' Perception of Climate Change

The present study asked farmers, "To what extent have you observed any enduring alterations in precipitation patterns within the last decade?" this inquiry aimed to gauge their perceptions on climate change. Previous studies in ethnographic research (West et al. 2008; Marin 2010) have shown that individuals possess the ability to accurately discern changes in climate patterns over a span of ten years via their personal encounters. Consequently, this study has used the mental map methodology to quantitatively assess farmers' observations and perceptions of climatic fluctuations. Once they affirmed their agreement, they were then inquired about the potential drop in rainfall amounts. The investigation also included the frequency with which farmers encountered adverse weather conditions and the subsequent impact on their agricultural yield during the preceding ten-year period. The study area has been found to experience of significant climatic events, namely drought in the Bundelkhand region and central region.

The results indicate that about 90.50% (highest) of farmers in Jhansi and 82.50% (lowest) perceived that rainfall has declined over past 10 years (Table 11.2). The majority of farmers perceived that rainfall distribution become erratic and unpredictable over past 10 years. In majority, farmers also perceived that temperature has increased and summer days are now hotter over past 10 years. The comparative analysis across the districts indicates the farmers in Jhansi district are well aware of changing rainfall and temperature patterns, while farmers in Barabanki are less aware. The results align with the data published by the Indian Meteorological Department, Government of India (2021). The temperature data for Uttar Pradesh indicate a notable upward trend in yearly temperature levels, with an average rise of around 0.01 °C per year seen between the years 1980 and 2020.

Table 11.2 Farmers' perception of changing climate

Indicators	Bundelkhand region		Central region	
	Jhansi	Lalitpur	Lucknow	Barabanki
Rainfall has declined over past 10 years	90.50	92.75	85.50	82.50
Drought frequency has increased over past 10 years	95.50	88.50	80.50	78.50
Summer days are become hotter over past 10 years	95.50	90.50	85.75	75.75
Water levels has declined over past 10 years	90.50	82.50	85.50	78.50
Weather extremes have become common phenomena over past 10 years	95.75	90.25	80.50	80.25
Late withdrawal of monsoon is now common phenomena	90.25	90.75	85.75	80.33
Rainfall distribution become erratic	92.75	90.25	89.25	82.25
Temperature has increased over past 10 years	95.25	92.50	90.27	89.25

Source Field Survey, 2022. *Note* values are in percentage

11.3.2 Adaptation Strategies in Surveyed Area

Table 11.3 presents the adaptation strategies adopted by sample farmers in both Bundelkhand and Central regions of Uttar Pradesh. The study has identified seven modes of incremental adaptations: (i) cropping pattern change, (ii) switch to non-farm activities, (iii) improved irrigation facilities, (iv) planted trees surrounded fields, (v) early maturing varieties, (vi) less water consuming crops, and (vii) crop diversification. Non-farm activities, improved irrigation facilities, and less water consumption were key adaptation strategies adopted by the farmers in Bundelkhand region. Planting trees surrounding the fields, using early maturing varieties of seeds and crop diversification were identified as the adaptation strategy for climate change in Central region. When faced with modification in planting operations, adjusting the sowing dates is often used as a technique. Typically, there is a self-regulated adjustment in altering the timing of planting activities in accordance with the commencement of monsoon season. Following the occurrence of crop failure during the Kharif season as a result of unpredictable rainfall patterns, it was noted that a subset of farmers opted to augment their agricultural intensity during the subsequent Rabi season. The farmers who encountered an early season setback proceeded to plant either the same or other crops in the same field. In order to save money, farmers reported hiring more members of their own families to work on their farms.

The farmers also reported to have made systemic adaptations by switching to early-maturing varieties of seeds and climate-resistant crop varieties. Different drought-tolerant and less water-consuming crops were favored by farmers in the rainfed (Bundelkhand region) and semi-arid (Central region) regions, where droughts are common. Farmers have shifted to high-yield, short-duration crop varieties to maximize profits. It was also observed that certain effective harvesting and water management measures were used, such as the use of drip irrigation, sprinklers, and the construction of rainwater harvesting structures. However, the extent to which these ideas were incorporated was modest. However, the farmers are myopic in their view of sustainable use of water, and they care exclusively for short-term profits. During the

Table 11.3 Adaptation strategies adopted by sample farmers

Indicators	Bundelkhand region		Central region	
	Jhansi	Lalitpur	Lucknow	Barabanki
Cropping pattern change	79.25	82.25	60.25	72.50
Switch to non-farm activities	50.25	55.75	80.25	42.25
Improved irrigation facilities	55.75	60.25	40.25	35.25
Planted trees surrounded fields	75.25	78.25	80.25	82.25
Early maturing varieties	80.75	75.27	76.50	70.25
Less water consuming crops	85.25	80.25	45.25	60.65
Crop diversification	45.25	42.25	45.75	48.25

Source Field Survey, 2022. *Note* values are in percent

field survey it was observed that many farmers were engaged in planting and selling eucalyptus timber even with the full knowledge that these trees are responsible for lowering the groundwater levels having negative impact on farming.

The other frequently documented systemic adaptation methods in farming practices were crop rotation, crop diversification, inter-cropping, and mixed cropping to enhance agricultural activity. According to Tripathi and Mishra (2017), diversification of crops in agricultural practices has been shown to reduce sample households' susceptibility to adverse weather conditions and unpredictable monsoon patterns. This is attributed to the provision of supplementary income opportunities compared to monoculture. The results from Table 11.3 indicate that farmers have adopted differential adaptation strategies to maximize farm returns across the districts. Nearly 80% of farmers in Jhansi district have changed their cropping pattern from water-intensive crops i.e., wheat to less water consuming crops like chickpeas to deal with climate change and save water, while only 60.25% of farmers in Lucknow district have changed their cropping pattern. Further, non-farm employment opportunities ensure regular income. As Lucknow district is the headquarters of Uttar Pradesh and has better non-farm employment opportunities, more than 80% of farmers have diversified their occupation patterns and switched to non-farm activities, while only 42.50% of farmers in Barabanki have diversified their occupation patterns. Improving irrigation systems along with conservation of water bodies ensures water even on hot summer days. It is observed that more than 60% of farmers in Lalitpur have improved their irrigation facilities, while only 35.25% of farmers in Barabanki have improved their irrigation facilities. Also, it is observed that more than 75% of farmers across the regions have planted Eucalyptus trees to maximize farm returns. More than 80% of farmers in Jhansi are concerned about water use in agriculture as the Bundelkhand region is facing continuous droughts. Hence, they have used early maturing and less water-consuming varieties to deal with the changing climate, while the corresponding figure for Barabanki district was only 70.25%. Lastly, crop diversification is the ultimate solution to combat the changing climate. It is observed that nearly half of the farmers have diversified their cropping patterns in the sample districts.

11.3.3 District-Wise Exposure Index

Farmers' perception of climate change was used to develop an exposure index which is a key part of livelihood vulnerability index. Further, exposure indices were calculated using Eqs. 1 and 2 (as explained in the previous section) for different districts of Bundelkhand and Central regions of Uttar Pradesh. The indicators mentioned in Table 11.1 were used to develop an exposure index for different districts. Questions were asked in a systematic manner to capture farmers' perceptions of climate change like, "Do you perceive that rainfall has declined over the past 10 years?". Then the rainfall indicator of the exposure index was calculated. The results show that Jhansi district was the most exposed district in the Bundelkhand region, while Lucknow was the most exposed district in the Central region (Table 11.4).

Table 11.4 District-wise exposure index in surveyed area

Indicators	Bundelkhand region		Central region	
	Jhansi	Lalitpur	Lucknow	Barabanki
Rainfall	0.91	0.87	0.77	0.76
Summer	0.81	0.79	0.79	0.79
Drought	0.87	0.81	0.90	0.83
Water level	0.84	0.88	0.78	0.83
Exposure index	0.86	0.83	0.81	0.80

Source Field Survey, 2022

Long summer days result in heatwaves and dry spells, which are responsible for drastic reductions in crop production. The calculated indices show that farmers belonging to the Jhansi district are highly exposed to the heatwaves and summer days in the Bundelkhand region, while mixed results were observed in districts in the central region. Further, drought is the most responsible factor for crop failure in both regions. The calculated index value shows that farmers (about 87.30%) in Jhansi district confirmed that frequencies of droughts have increased over the past 10 years, while 90% of farmers in Lucknow district also perceived that frequent drought incidence is responsible for higher climate exposure.

When districts experience continuous drought, longer summer periods, and erratic rainfall distribution patterns, it motivates the farmers to dig deeper bore wells to extract groundwater for irrigation and domestic purposes. This results in a further decline in water level. The calculated water level indices show that Lalitpur district in the Bundelkhand region is highly exposed among the districts in both regions. Finally, the calculated exposure index for all four districts and two regions shows that farmers in Jhansi district in the Bundelkhand region and Lucknow in the Central region are highly exposed to changing climates.

11.3.4 District-Wise Sensitivity Index

Using the socioeconomic data mentioned in Table 11.1, a sensitivity index was calculated for the surveyed districts. The calculated results show that farmers in Jhansi district in the Bundelkhand region and farmers in Lucknow district in the Central region were highly sensitive to changing climates (Table 11.5). Farmers are highly dependent on non-renewable forest resources for cooking, drinking untreated water, and living below the poverty line. Further, limited access to electricity and sanitation facilities is adding an additional layer of sensitivity to the system. The descriptive and cross-indicator analyses show that more than 90% of farmers in Jhansi districts are dependent on forest resources for cooking fuel, while only 61.80% of farmers in Barabanki are dependent on forest resources.

Table 11.5 District-wise sensitivity index in surveyed area

Indicators	Bundelkhand region		Central region	
	Jhansi	Lalitpur	Lucknow	Barabanki
Cooking source	0.91	0.80	0.76	0.61
Hand pump	0.92	0.77	0.56	0.43
Irrigation	0.89	0.56	0.47	0.32
Female-headed	0.83	0.74	0.62	0.40
Free medical facility	0.73	0.56	0.39	0.27
Sanitation facility	0.90	0.85	0.78	0.67
Nature of house	0.95	0.78	0.66	0.57
Below poverty line	0.92	0.87	0.75	0.59
Electricity access	0.93	0.89	0.81	0.77
Education level	0.96	0.92	0.86	0.80
Sensitivity index	0.89	0.77	0.66	0.54

Source Field Survey, 2022

Safe drinking water helps farmers deal with health-related issues. It is observed that 92% of farmers in Jhansi district used hand pumps for drinking water, while only 43.90% of farmers used hand pumps in Barabanki district. Nearly 90% of farmers are dependent on the government for irrigation, as the Bundelkhand region has the largest coverage of canal irrigation in Uttar Pradesh, while the corresponding figure for Barabanki was only 32.50%. It means farmers in Barabanki have the highest irrigation security due to higher coverage of irrigation and farmers in Barabanki have highest irrigation security. Likewise, more than 80% of households are headed by females in Jhansi district, while only 40% of households are headed by females in Barabanki district. In total, farmers belonging to Jhansi district are highly sensitive, while farmers belonging to Barabanki district are less sensitive among the surveyed districts.

11.3.5 District-Wise Adaptive Capacity Index

Changes in cropping patterns, diversification of crops, earlier planting and later harvesting, increased storage capacity, expert agricultural training, and water conservation are just some of the climate risk management strategies that farmers implemented in the sampled districts. Most efforts were made without official government involvement, although fertilizer and water supplies were two notable exceptions. Subsidized farm pond digging was linked to both drought mitigation and job creation efforts, thanks to the Horticulture Department's assistance with drip irrigation. However, due to economies of scale, these plans were usually successful for big farms (Jatav 2020).

The calculated adaptive capacity indices for different surveyed districts indicate that farmers belonging to the Barabanki district have the highest adaptive capacity, while those in the Jhansi district have the lowest adaptive capacity to deal with a changing climate (Table 11.6). The cross-indicator analysis shows that about 12% of farmers belonging to the Barabanki district have changed their cropping pattern, while only 2% of farmers belonging to the Jhansi district have changed their cropping pattern. Likewise, about 26% of farmers belonging to the Barabanki district have diversified their occupation patterns by involving themselves in non-farm activities, while only 10% of farmers belonging to the Jhansi district are involved in non-farm activities.

Further, joint family structures provide a safety net against any disaster, including climate change. A joint family structure also ensured regular unpaid family labor, which was always available for work in agriculture. It is observed that about 31% of farmers belonging to the Barabanki district lived in a joint family structure, while the corresponding figure for Jhansi district was only 6%. More than 25% of farmers in Barabanki district consulted with agricultural experts to deal with the climate, while the corresponding figure for Jhansi district was only 4%.

It was found that the farmers in Barabanki are highly motivated and aware of the conservation of water bodies. It is observed that more than 25% of farmers conserved water bodies, while the corresponding figure for Jhansi district was only 9%. Crop insurance is an ex-post adaptation strategy against a changing climate. It is observed that more than 30% of farmers in Barabanki district have taken crop insurance, while only 9% of farmers in Jhansi district have taken crop insurance. Similarly, more than 30% of farmers have storage capacity to store farm produce in Barabanki district, while only 13% of farmers in Jhansi district have storage capacity.

Table 11.6 District-wise adaptive capacity index in surveyed area

Indicators	Bundelkhand region		Central region	
	Jhansi	Lalitpur	Lucknow	Barabanki
Cropping pattern change	0.02	0.06	0.10	0.12
Switch to non-farm	0.10	0.12	0.23	0.26
Joint family	0.06	0.10	0.24	0.31
KKC	0.04	0.11	0.12	0.25
CWB	0.09	0.12	0.19	0.26
Crop insurance	0.09	0.19	0.20	0.31
Storage capacity	0.13	0.23	0.26	0.30
Training	0.21	0.24	0.29	0.37
Use of balance fertilizers	0.08	0.12	0.15	0.30
Sowing dates	0.12	0.21	0.30	0.33
Crop diversification	0.08	0.18	0.26	0.32
Adaptive capacity index	0.09	0.15	0.21	0.28

Source Field *Survey*, 2022

Furthermore, about 37% of farmers have taken agriculture training at Kisan Vikas Kendra about modern tools and techniques in agriculture, while the corresponding figure for Jhansi district was only 20.80%. Balanced use of fertilizers is also important for sustainable farming. It is observed that about 30% of farmers in Barabanki district used the recommended fertilizer ratio, while only 8% of farmers in Jhansi district did so. More than 30% of farmers in Barabanki have adjusted sowing dates to deal with climate change, while the corresponding figure is only 12% in Jhansi district. In total, farmers belonging to the Barabanki district have the highest adaptive capacity, while those in the Jhansi district have the lowest adaptive capacity to deal with a changing climate.

11.3.6 District-Wise Livelihood Vulnerability Index

Using Eq. 6, the potential livelihood vulnerability index for different districts in Bundelkhand and the Central regions was calculated. The calculated results show that farmers in the Bundelkhand region were relatively more exposed and sensitive to changing climates compared with the districts of the Central region (Table 11.7). On the contrary, farmers belonging to the Bundelkhand region had the least adaptive capacity compared with farmers in the Central region. In total, farmers in the Bundelkhand region are relatively more vulnerable than farmers in the Central Region. The main reason of higher livelihood vulnerability in the Bundelkhand region was correlated with their lowest adaptive capacity. Lastly, among the four surveyed districts and 480 sample farmers, farmers belonging to the Jhansi district were relatively highly exposed, sensitive, and had the least adaptive capacity to deal with changing climates, whereas farmers belonging to the Barabanki district were relatively less exposed, sensitive, and had a higher adaptive capacity. In other words, Barabanki district was the least vulnerable to a changing climate among the surveyed districts.

Table 11.7 District-wise potential and livelihood vulnerability index

Indicators	Bundelkhand region		Central region	
	Jhansi	Lalitpur	Lucknow	Barabanki
Exposure index	0.86	0.84	0.81	0.80
Sensitivity index	0.89	0.77	0.67	0.55
Adaptive capacity index	0.09	0.15	0.21	0.28
Potential livelihood vulnerability index	1.75	1.61	1.48	1.35
Livelihood vulnerability index	0.68	0.53	0.40	0.28

Source Field Survey, 2022

11.4 Conclusion and Policy Recommendations

Farmers in Bundelkhand and the Central region, which are recognized as economically disadvantaged and environmentally susceptible areas, have expressed their apprehensions on the repercussions of climate change. Farmers held the belief that climate change poses a significant challenge to agriculture and the livelihoods of farmers. They expressed a desire to engage in partnerships with the government and non-governmental organizations (NGOs) in order to effectively adapt and reduce their vulnerability to the impacts of climate change. It was found that making use of extension services and having a higher level of education (above from secondary) help farmers to deal with changing climate. Further, higher income levels were the most significant economic indicator in minimizing sensitivity to climate change and its consequences on livelihoods. Farmers have been taking measures to adapt to climate change in the study area by installing new irrigation systems, shifting planting dates, altering their cropping patterns, diversifying their crops, and using less water-intensive seed varieties.

Therefore, the present study asserts that farmers, particularly those residing in developing countries such as India, must undertake measures to adjust to climate change with an aim to alleviate its adverse impacts and capitalize on its potential benefits. Now is the moment to help farmers to better understand climate change adaptation and to implement effective adaptation methods to mitigate its negative effects. The farmers' adaptability might be improved by the establishment of training programs, skill development, and capacity building. However, the government authorities need information from the field to devise appropriate programs and suitable implementation to provide farmers the right kind of technical assistance. Policymakers ought to strategically devise and implement suitable adaptation strategies to mitigate the adverse consequences of climate change. These strategies may encompass technological advancements aimed at enhancing irrigation and weather forecasting systems, cost reduction of agricultural inputs, facilitation of information, availability, provision of agricultural subsidies to farmers, and augmentation of access to agricultural markets. This imperative arises from the fact that farmers encounter numerous obstacles in their efforts to adapt to climate change that has been established by this study.

References

Abid M, Scheffran J, Schneider UA, Ashfaq M (2015) Farmers' perceptions of and adaptation strategies to climate change and their determinants: The case of Punjab province. Pakistan. Earth System Dynamics 6(1):225–243

Alam GM, Khorshal A, Shahbaz M (2017) Climate change perceptions and local adaptation strategies of hazard-prone rural households in Bangladesh. Clim Risk Manag 17:52–63

Barnett BJ, Olivier M (2007) Weather index insurance for agriculture and rural areas in lower-income countries. Am J Agr Econ 89(5):1241–1247

Bryan E, Deressa TT, Gbetibouo GA, Ringler C (2009) Adaptation to climate change in Ethiopia and South Africa: options and constraints. Environment and Science Policy 12(4):413–426
Deressa TT, Hassan R, Ringler C, Alemu T, Mahmud Y (2009) Determinants of farmers' choice of adaptation methods to climate change in the Nile Basin of Ethiopia. Glob Environ Chang 19(2):248–255
Fellmann T (2012) The assessment of the socio-economic impacts of climate change at household level and policy implications. Building resilience for adaptation to climate change in the agriculture sector. Agriculture and Food Organization, Rome, pp 37–61
Funk C, Raghavan Sathyan A, Winker P, Breuer L (2019) Climate change-changing livelihood: Smallholder's perceptions and adaptation strategies. Journal of Environment and Management 261:55–65
Hahn MB, Riederer AM, Foster SO (2009) The livelihood vulnerability index: A pragmatic approach to assessing risks from climate variability and change – A case study in Mozambique. Glob Environ Chang 19(1):74–88
Hoque MZ, Cui S, Xu L, Ismal I, Tang J, Ding S (2019) Assessing agricultural livelihood vulnerability to climate change in coastal Bangladesh. Int J Environ Res Public Health 16(22):4552
IMD (2021). Annual Report, Indian Meteorological Department of India. Ministry of Earth Science, Government of India, New Delhi, India.
IPCC, (2001). Climate change 2001: impacts, adaptation and vulnerability, Contribution of Working Group II to the Third Assessment Report of the Intergovernmental Panel on Climate Change, McCarthy, J.J., Canziani, O.F., Leary, N.A., Dokken, D.J., White, K.S. (Eds.). Cambridge University Press, Cambridge, UK, and New York, U.S.A.
IPCC, (2007). Climate Change 2007: impact, adaptation and vulnerability, Contribution of Working Group II to the Fourth Assessment Report of the Intergovernmental Panel on Climate Change, Martin, P., Osvaldo, C., Jean, P., Paul, V.D.L., Clair, H. (Eds.), Cambridge University Press, Cambridge, UK, and New York, U.S.A.
IPCC (2014). Climate Change 2014: Synthesis Report. Intergovernmental Panel on Climate Change (IPCC), Geneva, Switzerland.
IPCC. (2018). Summary for Policymakers. In: Global Warming of 1.5°C. Masson-Delmotte, V., P. Zhai, H.-O. Pörtner, D. Roberts, J. Skea, P.R. Shukla, A. Pirani, W. Moufouma- Okia, C. Péan, R. Pidcock, S. Connors, J.B.R. Matthews, Y. Chen, X. Zhou, M.I. Gomis, E. Lonnoy, T. Maycock, M. Tignor, and T. Waterfield (Eds.)], World Meteorological Organization, Geneva, Switzerland.
Jatav SS (2020) Bridging the gap between biophysical and social vulnerability in rural India: A community Livelihood Vulnerability Approach. Area Development and Policy 5(4):390–411
Jatav SS (2021) Does India achieve agenda 2030 targets: a multiple lens analysis? Indian Journal of Ecology. 48:1230–1245
Jatav SS, Kumar A, Malik BB (2021a) Impact of Covid-19 on the livelihood of rural farmers in Uttar Pradesh. India. Journal of Rural Development. 40(1):94–11
Jatav SS, Surendra M, Sanatan N, Sonali N (2021b) Coping to Covid-19 in Uttar Pradesh, India: evidence from NSSO 76th Round Data. Curr Urban Stud 9:206–217
Jatav SS (2022) Development of multidimensional food security index for Rajasthan, India: A district-level analysis. Local Development & Society 3(2):1–23
Kelly PM, Adger WN (2000) Theory and Practice in assessing vulnerability to climate change and facilitating adaptation. Clim Change 47(4):325–352
Kuchimanchi BR, Nazareth D, Bendapudi R, Awasthi S, D'Souza M (2019) Assessing differential vulnerability of communities in the agrarian context in tow districts of Maharashtra. India. Climate and Development 11(10):918–929
Kumar MD, Reddy VR, Narayanamoorthy A, Bassi N, James AJ (2018) Rainfed areas: poor definition and flawed solutions. Int J Water Resour Dev 34(2):278–291
Marin A (2010) Riders under storms: contributions of nomadic herders' observations to analysing climate change in Mongolia. Glob Environ Chang 20(1):162–176

Mertz O, Halsnæs K, Olesen JE, Rasmussen K (2009) Adaptation to climate change in developing countries. Environ Manage 43(5):743–752
Maddison, D. (2007).The Perception of and Adaptation to Climate Change in Africa. The World Bank.
Masud MM, Azam MN, Mohiuddin M, Banna H, Akhtar R, Alam ASAF, Begum H (2017) Adaptation barriers and strategies towards climate change: Challenges in the agricultural section. J Clean Prod 156:698–706
Miranda L, Hordijk M, Molina RKT (2011) Water governance key approaches: An analytical framework. Literature Review, Change & Sustain 4:1–23
Nadeem, S., Elahi, I., Hadi, A., & Uddin, I. (2009). Traditional knowledge and local institutions support adaptation to water-induced hazards in Chitral, Pakistan. Kathmandu, Nepal: The International Centre for Integrated Mountain Development (ICIMOD).
Nhemachena, C. & Hassan, R. (2008). Micro-level analysis of farmer's adaptation climate change in Southern Africa. International Food Policy Research Institute. 15(7).
Omerkhil N, Tara., Ch., Donatella, V, Alatalo, J. M., & Rajiv, P. (2020) Climate change vulnerability and adaptation strategies for smallholder farmers in Yangi Qala district, Takhar, Afghanistan. Ecol Ind 110:1–10
Rai A, Sharma SD, Sahoo PM, Malhotra PK (2008) Development of livelihood index for different agro-climatic zones in India. Agricultural Economic Research Review 21:173–182
Sanatan N, Surendra S (2020) Development of gender sensitive vulnerability index for urban Indian Slums. Indian Journal of Traditional Knowledge. 19(3):592–597
Singh S, Nayak S (2014) Climate change and agriculture production in India. Eur Acad Res 2:12–30
Singh S, Nayak S (2018a) Land inequality and agricultural sustainability in Uttar Pradesh, India: a regional analysis. Asian Journal of Science and Technology. 9:12–32
Singh S, Nayak S (2018b) Application of indicators for identifying climate vulnerable areas in sub-tropical regions of India. Journal of Multidimensional Research. 7:1–514
Singh NP, Surendra S, Bhawna A, Bal SK (2019) Climate vulnerability assessment in semi-arid and arid region of Rajasthan, India: An enquiry into the disadvantaged districts. Journal of Agrometeorology 21(2):197–202
Singh S, Sanatan N (2020) Development of sustainable livelihood security index for different agro-climatic zones of Uttar Pradesh. J Rural Dev 39(1):110–129
Singh S (2020) Farmers' perception of climate change and adaptation decisions: a micro-level analysis of farmers in Bundelkhand Region. Ecol Ind 116:106475
Shrestha R, Chaweewan N, Arunyawat S (2017) Adaptation to climate change by rural ethnic communities of Northern Thailand. Climate 5:57
Sikka A, Islam A, Rao KV (2018) Climate-smart land and water management for sustainable agriculture. Irrig Drain 67(1):72–81
Tripathi A, Ashok M (2017) Knowledge and passive adaptation to climate change: an example from Indian farmers. Clim Risk Manag 16:195–207
West CT, Roncoli C, Ouattara F (2008) Local perception and regional climate trends on the central plateau of Burkina Faso. Land Degrad Dev 19:289–304
World Bank (1997) World Development Report 1997: The state in a changing world. Oxford University Press, New York, NY
World Meteorological Department (WMO). Global temperatures set to reach new records in next five years. Access on 26.09.2023. https://onlinelibrary.wiley.com/doi/10.1002/ldr.842

Chapter 12
Religion as a Means to Address Disaster Uncertainty: Case Study of Kullu and Mandi District, Himachal Pradesh

Katyayini Sood

Abstract Throughout human history, numerous disasters have plagued humanity, and various religions worldwide have attempted to explain their occurrence through diverse mythologies and folklore. However, very little is known about the fascinating *deo/devta/deota* tradition in the Kullu district of Himachal Pradesh, also called the *devta* tradition or local deity tradition. This chapter sheds light on this unique tradition, in which every village in the district has its deity believed to protect its inhabitants from disasters and uncertainties. This study delves into religion's role in reducing or eliminating the uncertainties caused by disasters. The study highlights various rituals and practices, highlighting people's reliance on cultural values and traditions. It emphasizes the importance of local beliefs and practices when dealing with disasters and how they can provide security and reassurance to those affected. The local deities, who are considered the guardians of their respective villages or regions, assist people in coping with disaster-related uncertainties through the use of shamans. According to local beliefs, the deities can also cause disasters in the valley if they become angry.

Keywords Disaster uncertainty · Risk perception · Shamanic tradition · Culture · Religion

12.1 Introduction

The Oxford Dictionary describes uncertainty as something which is not definite or not determined. The term uncertainty is also well suited in disaster studies because a few events like earthquakes and tsunamis are hard to determine. The exact time and location of the disaster is something humankind is always unsure about. Uncertainty is not merely limited to just one phase of disaster but exists in almost all the phases of disaster. The phase of uncertainty does not end here but continues till the

K. Sood (✉)
United Nations University, Bonn, Germany
e-mail: sood@ehs.unu.edu

A. Sarkar et al. (eds.), *Risk, Uncertainty and Maladaptation to Climate Change*,
Disaster Risk Reduction, https://doi.org/10.1007/978-981-99-9474-8_12

recovery phase. In the cascading disaster events, the uncertainty further rises. Beck (2010) has reported a rise in man-made disasters, and Bloom et al. (2018) support that technological advancement is the root cause of increasing uncertainty. Rocha et al. (2009) elaborate on how disasters are uncertain events as there is a lack of security concerning time, place, site, periodicity, circumstances, magnitude, information, and knowledge. The prolonged uncertainty leads to a slower recovery rate, according to Jovanovic and Ma (2020). Uncertainty in disaster management is also unique as providing humanitarian assistance becomes difficult as it deals with uncertainty in demand, supply, and even the location of the place where the assistance is needed (Liberatore et al. 2013). Uncertainties overlap and are more visible in the vulnerable population (Parthasarathy 2018). Uncertainty is multidimensional and impacts human settlements, ecology, and the countries' economy. As the uncertainty rises, it is anticipated that events like floods, drought, etc., will be of higher magnitude (Avery 1998). According to a report by AIDMI (2018), the levels of uncertainty have been rising due to climate change caused by anthropogenic activities.

There is a distinction between risk and uncertainty. As for risk, the loss can be anticipated, but nothing can be determined regarding uncertainty. There are many possibilities of something happening, but what will happen eventually remains a question. Schnarr and Mertz (2022) state that uncertainty cannot be converted entirely into risks because a person can never know everything. He emphasizes the difference between the two and says uncertainty is a broad possible outcome. There are too many, so it cannot be determined which will turn out to be accurate, whereas, in risk, he says that the possibility of effect and exposure is determined. Culture plays a vital role in addressing risk as well as in uncertainty. Right from the very beginning, everyone adopts some cultural traits. Hence, these cultural traits help us identify how and why a person reacted to some situation in a particular way. Culture identifies some reasons or logic behind all the activities done. Douglas (1966) also proposed the concept of purity and danger. What is dangerous or "not pure" is termed as risk. Douglas (1966) believes every person has some different perception of risk according to the culture people inherit. Every individual decides what he should fear and what is in the capacity to control and hence what one should not fear. Douglas (1966) focuses on the politicization of the concept and urges people to ask who is to be blamed. The cultural theory posits that risk perception is a "culturally standardized response". Sherry and Curtis (2017) depict that even during the scientific revolution, we have found scientific evidence to prove why catastrophes occur. However, people still align themselves with religious beliefs and thoughts. When these disasters happen, people tend to believe in the supreme being created by humans. Undoubtedly, religion is one of the components of culture and is the most critical institution in all societies across the globe. Every eighth person out of 10 identifies as part of a religion (Hacket and Grim 2012). Bentzen (2019) states that disasters come with a lot of uncertainty and shocks to adversities; in these uncertainties, the sense of belonging to a religion strengthens. Religion is understood as a set of particular beliefs which people tend to believe, mainly the idea of God or a supernatural power that runs the world. Disasters in primitive societies were understood as God's way of showing discontent. Thus, people personified everything they feared and started worshipping them, and as time passed, various myths and folklore came into existence. Religion is a part of culture; folktales and myths are components of religion. Mythology is the stories that revolve around 'God' and provide a reason to believe in a religion. Folktales, on the other hand, not just talk about legends but also socio-cultural practices and social mobility.

Religion plays a central role in all the phases of the disaster management cycle, from preparedness to recovery (Regad and Silva 2020). It governs the way of life, how individuals perceive and react to disasters and even unites people during disasters. Almost all texts of all religions talk about disasters. Even in contemporary societies, where technology has replaced almost everything that existed in the past, the idea of 'God' remains untouched. In recent years, the concept of religion has gained importance in disaster studies as some folktales and myths have predicted disasters, and some suggest mitigation strategies to avoid uncertain situations. In some ways, all religions have focused on some aspect of disaster through texts, folktales, or hymns, albeit to different extents. A few religions emphasize how disasters result from disturbing the equilibrium in nature.

Mumru and Pramanik (2018) assert that the *devta/deota/deo* culture in the upper regions of Himachal Pradesh is another example of this type of religion. Some call it a form of 'shamanism,' while others call it 'animism.' The culture revolves around local deities who move on palanquins or raths to ensure the community's well-being. Each village has its deity who protects them from disasters and uncertain events. This tradition acts as a form of governance, with the *Shaman (gur)* serving as the messenger between the people and the gods. While religion plays a significant role in administration, there is little literature on mainstream religion's involvement in public administration.

Chunhabunyatip et al. (2018) note that spirits and gods play a crucial role in preserving the resources of Indigenous communities. These traditions are old but significant. Bentzen (2019) states that once the philosophers suggested that religion would end soon in modern society, its importance has not withered away. It is in the time of disaster that people associate themselves with 'God' or any religion; during the disaster, google searches for God and his prayers increase. There is a theory of religious coping by Pregament (2004) in which he advocates how religion helps in times of crisis in one's life. Religion is a mechanism that helps one deal with life's uncertainty. It is not in all the disasters that the uncertainty increases but in the ones which are dreadful enough to disturb the people emotionally and cause stress among them. Bentzen (2019) gives examples of earthquakes, tsunamis, and volcanic eruptions, which are relatively uncertain and where religiosity increases. In contrast, religious dependence does not see a rise in cyclones that are very much anticipated.

Emotion-focused coping is essential in the recovery phase; religion becomes a pillar. Each society has two types of institutions, formal and informal. The formal institutions are impersonal and follow the rules of the law more in terms of the written form. The informal institutions are more personal and lack a written set of rules and laws. They are culturally very significant but lack transparency (Henriks 2010). The two co-exist in all civilizations. For inclusive governance, it is essential to consider both institutions to remove social disparity and exclusion. Humans are culture-conditioned beings who make decisions determined by their cultural environments (Misiura and Rozkwitalska 2019). Chotkowski (2010) states that these institutions form the base of human behavior. Brännmark (2018) "Institutions are the game principles in society, and they come as limitations that people develop to form political, economic, and social interactions. They include informal limitations, namely sanctions, taboos, customs, tradition or principles of behaviour and formal principles, such as constitutions, laws, and property rights." Thus, it can be anticipated that both formal and informal institutions play an essential role in the field of disaster management as well.

The chapter explores the myths (beliefs which are based on tradition), mythologies (stories which involve a supreme power or supernatural aspect), folktales and the shamanic tradition or the local deity institution concerning disaster uncertainty by engaging with the community which functions under the leadership of *God Raghunath* and *Hurung Narayan.* The study focuses on exploring the understanding and perception of disaster uncertainty through mythological stories and folktales. It addresses research questions like, how does the local deity signal the community under his authority to escape a disaster, and how do people interpret the state of trance in *gur*? What do people do to eliminate uncertainty? How are disaster uncertainty and risks perceived? It also highlights the obstacles the formal institution faces in implementing its policies.

12.2 Materials and Methods

The study is based on ethnographic methods. Ethnography entails a long period of observation of a group while the researcher is involved in their everyday routines. For this research, people under the leadership of Raghunath of Kullu Valley and Hurung Narayan in Chuhar Valley were observed. The primary sources included interviews with participants and photographs and videos taken during data collection. Purposive sampling was used to interview the five individuals closest to God to understand the relationship between shamanic traditions and disasters. For formal authorities, convenient sampling was used, as it depended on the availability of officials. A maximum variation sample was used to obtain diverse information regarding disasters and the local deity system. This included individuals from different genders, classes, castes, and educational backgrounds. Initially, the sample size was not predetermined. Data collection ceased once data saturation was achieved. A total of 40 people were interviewed, including six individuals who lived near the deity and six government officials from different departments associated with disasters.

The data was collected through unstructured interviews with three groups: local devotees, the local administration, and those knowledgeable about the deity system's relation to disasters. Additionally, data was collected through observation and audio-visual recordings taken during the fieldwork. Various photographs and videos were also collected during events, including *Jagatis*, to discuss solutions to extreme rain in July 2021. Additionally, secondary sources included past documentaries highlighting the importance of deities in the lives of locals in the Kullu and Mandi districts. These documentaries were filmed during local fairs, which hold significant value in the community. The venue varied for all the interviews as some incidents or observations could occur at a particular location. It varied from the house to the Jagati path in Nagar.

12.3 Findings

12.3.1 Schein's Multi-layered Organizational Culture Mode

Schein's cultural model helps us understand the relationship between culture and disasters in a better way. Schein (2012) defines culture as a pattern of basic assumptions, (2) invented, discovered, or developed by a given group, (3) as it learns to cope with its problems of external adaptation and internal integration, (4) that has worked well enough to be considered valid and, therefore (5) is to be taught to new members as the (6) correct way to perceive, think, and feel about those problems. This definition can be looked upon from the context of a disaster as what are the basic assumptions of the group concerning disasters and how the group forms its ways of fighting the external elements (disasters). When these ways (rituals) work, they are carried forward from one generation to another; with the help of verbal communication, people make a set of norms to teach future generations on how to react to disasters.

Hebb (1954) produced an analysis of how the set of assumptions that a group follows lowers anxiety levels among people. Cultural beliefs are ubiquitous, and although they vary from one society to another, the aim is to provide people with some comforting measures during uncertainty. Schein (2012) states that each culture takes time to be formulated. It is not an instant process. The ancestors or seniors of a society evaluate all the regulations. He then defines three levels of culture and justifies himself by stating the need to differentiate between the different levels of this culture that consists of certain visible aspects of culture and many underlying factors that support its existence.

12.3.2 Explanation of the Layers in Disaster Context Underlying Assumptions

The underlying assumptions are the founding stones of any culture. These assumptions have passed the test of time and have been proven successfully over the years. This further becomes generalized in a way that the people of society unconsciously follow them. These assumptions become the base of the values and actions of an individual. Regarding disasters and deities' relation, the basic assumption is that God is the ultimate owner of the resources, and humans are merely using them for the time being. The residents believe that 'God' himself owns the natural resources they use. Hurung Narayan owns vast tracts of forest land, and the people in his authority can take the wood only after the locals ask for permission, as folklore revolves around this value.

Once upon a time, a person took wood from this forest to build his house, failing to realize that taking wood without his permission is a sin, later, he had serious health problems and was trapped under massive debt. While narrating this the Gur said even if you take this wood from these forests, you will never be happy, it will be a curse for life. So, people for ages have been taking permission to cut trees and collect wood from the forest.

Another underlying assumption is that whatever is offered to the good should be pure and given with the right hand instead of the left. In Hinduism, all religious activities, like offering prayers, rituals, and eating, are preferred to be done with the right hand. These assumptions may or may not have proper justification but are accepted by everyone in that group. These assumptions are nearly impossible to change. This is because they are very deep-rooted in society. The examples mentioned above are also not subjected to any debate, like people belonging to this tradition do not question land ownership or debate on why one cannot use the left hand for performing any rituals. This makes them self-evident, and thus, people follow them unconsciously. These include thoughts and feelings that the people sharing these cultural values have.

During times of disaster, a fatalistic attitude among the elderly often develops due to the pervasive belief that all events are the result of divine will. This attitude can have significant consequences, as it may lead to limited actions being undertaken during the preparedness phase. As a result, it is crucial to recognize and address the impact of religious assumptions on disaster preparedness and response efforts, particularly among older adults. By taking proactive measures to address these issues, we can help ensure that our communities are better equipped to respond to emergencies and minimize the impact of disasters on vulnerable populations.

Values are related to the code of conduct. The statements or rules given by the institution are included in this. They shape the way of life, and over the years, they have also become an essential aspect of the culture. They have a goal and a philosophy behind it, including people's beliefs. The values regarding disasters can be looked upon with the help of the following example.

People in the region offer their harvest, milk, or any other produce to these deities. The objective is to show respect towards the deity. The philosophy of giving to God is based on the fact that "You give 10 he turns in 100. He doesn't keep anything to himself but returns double of everything you offer him". People offer these to protect the source of their livelihoods. This ritual is associated to being safe during calamities and secure livelihoods for assuring their financial stability.

People stay away from the temple who have witnessed a death or childbirth. This period is generally associated with impurity also knowns as Sutak in local language. Polluting the temple premises is seen as an act which causes anger among the deities. This may result in abrupt or heavy rainfall which cause damage to the crops.

The values are practiced at an individual level. The group's values are stringent. Although, with time, some anomalies are witnessed. Nevertheless, the value system tries to inculcate some good practices in the life of people. These include respecting natural resources, not overexploiting or polluting the waters, etc. These value systems help in environmental protection. Moreover, proposes disaster risk reduction methods from an Indigenous knowledge perspective. Researchers are proposing methods to incorporate these value systems and rituals at more significant levels to contribute at a much larger spatial scale.

Dev Khel is a local ritual where the *gur* dances before the deities and goes into a trance to communicate with the Gods. The *gur*, advised by the gods then suggests rituals that must be followed to stop disasters from happening, and even if they occur, their impact is significantly less. This is believed to be an ancient practice performed when kings visited the deity occasionally to ensure stability in their kingdom. One story reported by the local people in the Mandi district is regarding Dev Aadi Brahma, one of the superior deities of Mandi. Once, there was a spread of communicable disease in the province. During that time, he provided ash and asked the king to spread it across the kingdom. According to the local folklore, this ash worked as a fence from external evil powers. After this, spreading the ash across Mandi town became a tradition during the *Shivratri* fair. This, even today, according to locals, protects them from evil powers which have the potential to harm them. In the local language, it is known as *Suraksha kawach* (shield against bad).

Still today, during the ritual of *dev khel*, the *gurs* communicate with the god to predict disasters and advice local people to perform a pooja or sacrifice to eliminate the loss of life. During COVID-19, there was a state of panic. Most of the villagers went to the local *devtas*. There were gatherings to ask the *devta* what would have happened next. How do they eliminate the risks and other threats? Most deities asked the people to remain calm and said they would protect them from the possible consequences of disasters. Although people were asked not to gather for meetings by the formal institution, they did participate in the prayers, thinking it would stop the spread in their areas. According to locals, the gods' anger led to a surge in the number of cases as they had failed to perform the annual functions of these gods.

Havans are conducted at individual and community levels in reverence to the gods and the *yoginis* in which people believe. In Shikari devi in Mandi special *havan* takes place to bring protection to lives from disasters. *Paath,* in general terms, is the regular prayer people offer on special occasions. People believe in the *paaths* because it provides them with psychological support during times of disaster. Some *paaths* are short, whereas others (like *Maha Martyunjaya*), may take 3–7 days. These *paaths* provide local people with religious perspective and solace to accept risk. A woman respondent who had recently conducted a *paath* at her home to address the problem of low crop productivity commented, "*The priest asked her to conduct a Gayatri paath. I do not know if things will be fine, but my attitude will change. My attitude towards looking for things might change. I hope I will not take everything as a disappointment. I will think I have given something, and he (God) will return me more than I have.*"

Animal sacrifice is a common practice in the region. It is not seen as a violation of animal rights but as an act one should purposely do when one desires something. Almost all the gods accept it as a gift, and usually a goat is given as a sacrifice to please God to bless local people with safe conditions with enough produce, comfortable living, etc. There are unique places where the offerings are made. The quality of the goat should also be good; otherwise, when these deities feel aggrieved, they purposely create situations in which people go back to them and then apologize. The irony is that although the deity looks after most of the ecology, the gods still accept the animal sacrifices.

Jagatti Patt is a famous temple that is made of stone and wood. According to the local beliefs, the stone slab inside the temple was brought in by the honeybees from Deo Tibba, a mountain in the Pir Pranjal Range that was considered as the home of gods. During the olden times, it was a common place where all the people and their deities gathered whenever natural calamities like floods and earthquakes occurred. The processions are called *Jagatis* and are called whenever gods feel that there is a presence of hazards. The information of the *Jagati* is circulated within the community.

The two most famous dev melas are the *Shivratri mela* in Mandi district and the Dussehra mela in Kullu district. These two are essentially crucial in order to understand their effect on COVID. In 2020, listening to the guidelines, it was proposed that these gatherings should not be held during the outbreak. The District Collector formally invites the deities and asks them to participate in this fair. However, only a few prominent deities were invited to Kullu in 2020, followed by a *jaggati.* The deities warned the authorities through their *fi* not to repeat the same. They highlighted how popular culture could not suppress the local culture because it can be harmful. Later in the year 2021, all the deities were asked to be a part of the culture, as the officials had apologized for it last year.

12.3.3 Case Study of Malana

Malana, a village in the Kullu district, is known to be the oldest democracy in the world. Located at 8700 feet, the village lies near Nagar in Parvati Valley. The total population of the village is nearly 2300, which makes it one of the most populated villages of the valley. It is one of the secluded places of the region; this is attributed to the deity that rules the region. The chief deity of the region is *Jamdagni, Maharishi* (*Jamlu Rishi*). The mythological lore around this deity is that he did penance and secured Shiva's blessings. When asked what he wanted in return, he relied on the fact that he wanted a place with abundant nature. While moving to this place, he observed that his brothers were following him. He wanted to be alone, so with his supernatural powers, he covered the area with mist. His brothers lost track and took different paths. One ended up in Banjar Valley, the other went to Lahaul Spiti. *Jamlu Maharshi* wanted to establish his kingdom in Malana, but it was ruled by a demon known as Banasura. They both fought for supremacy, and the fight ended on mutual terms.

However, as time passed, *Jamlu Devta* became superior and gained societal recognition. The traditions people followed after the agreement are assumed to be the same. Society is very rigid in terms of rules and regulations. The representatives are still elected, and the deity makes the critical decisions of the villages. The village is also known as "Touch Me, not Village." The tourists cannot touch any village property or enter the courtyards. If they purchase anything, it has to be kept at the counter by the shopkeeper and the money is also kept at the counter so that there is no direct contact between the two. If a villager is accidentally touched, they rush to take a bath as it would remove the impurities from the person and the tourist, or the outsider has to pay the fine to the village administration. Until the 1980s, the village had been in isolation. This was due to the remote geographical location of the village. Another reason for the village's uniqueness can be attributed to the people's religious devotion toward their deity.

From 1970 onwards, the tourists started flowing in and taught people to extract hash from the plants. Years later now, the area is known for the best hash it provides to tourists. When tourism started taking over the belief system and traditions of the village, the deity declared the closure of rest houses in the village for a while. People respected his orders and later convinced him that this was a source of livelihood for them. He permitted it on the condition that people would not forget their traditions. The faith in the god leads to respect. People respect him and thus follow his instructions. They fear his anger as the locals firmly believe nothing remains hidden from God. The feeling of being watched is never out of the way. If one feels his problem is not being solved, he can seek help from the formal institutions, but in return, he has to submit 2500 rupees to the local council. This is seen as a punishment for going against the deity. Only a few people opt for this method of justice. Looking from a disaster point of view, being located in a highly hazardous location, the village has been safe from all kinds of disasters. According to a local, till now, none of the people has lost their lives because of disasters. Their deity protects them from catastrophes. This is because they follow the guidelines and follow a way of life as ordered by the deity. One of the most unique phenomena was observed during the coronavirus when the entire nation was in a crisis, this village remained untouched by its effects.

12.3.4 COVID-19 Interventions

As per the interviews and newspapers, there was a claim that the deity had warned the local people regarding the upcoming biological disaster. The lockdown was imposed before the national lockdown, and as people became aware, they went to the deity to protect them from upcoming perils. The deity assured them he would protect the community from the biological disaster. He asked the village residents to close the village border to outsiders for preventive measures. The people had isolated themselves from the normal public of the region. According to the locals, this was the main reason they were free from the disease for a long time. However, the residents refused to wear masks and followed social distancing. The authorities had been concerned about the outbreak in the resource deficit region. The formal institutions believed that social distancing was not possible because of the social customs and traditions which are followed in the region. The council's norms formed at the village level included a fine of rupees fifty-one thousand (613 USD) on anyone who had violated the rules and regulations to protect the village from spreading the coronavirus for the first year (2020). However, as the covid restrictions eased, the villagers were allowed to put up tourist camps outside the village. These strategies are imposed by the locals with the guidance of the deity.

The residents did not go for testing initially. They firmly believed they were being protected against the disease by the god. They attended frequent gatherings and religious festivals like the *Fagli,* a five-day festival for the deity himself. The entire village, including children, participated.

12.3.5 Vaccination Policy

The Government of India started a vaccination drive within the region to prevent the worsening of the virus. The residents of this village refused to take the vaccination, stating that their deity did not allow them to take it. This was not the first time that people refused to take vaccines. It was only after immense efforts of the authorities that the residents agreed to provide vaccination to their infants. The primary concern of people was that the vaccine might not suit them. They saw it as a threat to progeny, as they do not marry within other villages, believing it would harm their existence as a village. These statements hindered the state in achieving 100% vaccination status.

Despite being aware of the deaths happening in the second wave, people did not agree to vaccinate themselves. The deity was entirely against this vaccination drive which was happening. It was a serious effort of the administration; the District Commissioner, along with the health department, had to visit the village repeatedly, to convince the village parliament and request the local deity. Only in September that year the villagers, with the permission of their deity started vaccination process which was much later than what other parts of the country were experiencing.

There is a lot that can be comprehended from the instances above. People have a strong sense of attachment to their deity. They respect him, and also, they are afraid of his anger. Although the deity is also seen as protective when people do not follow his instructions, he becomes angry, and the community has to fulfill his demand to pay a fine. It is the respect toward the deity that people rarely go against him. They see him as the ultimate power who will provide them with solace and protection. So, there is trust, which leads to respect, and it is because of respect that they fear the deity and follow whatever he says.

12.4 Conclusion

In today's world, disasters are becoming increasingly frequent, leading to greater risks and uncertainties. The state of Himachal Pradesh is susceptible to 25 different types of hazards, including flash floods, landslides, and snowstorms, each with varying degrees of intensity. Settlement patterns are scattered and isolated, and the complex geography ranges from 600 to 6000 m in elevation. Recent disasters have revealed poor disaster risk reduction measures and early warning systems in place. For a long time, disasters were attributed to natural phenomena caused by divine anger. In Himachal Pradesh, local deities play a crucial role in eliminating disaster uncertainties and trust in the local deity system is strong. However, this trust can become a hurdle when the administration needs to act to reduce people's vulnerability.

Culture shapes people's values and beliefs, influencing their perspectives on life and disasters. Even in today's technology-driven world, people often turn to religion when dealing with these uncertainties. Religion provides some benefits to society and remains a crucial factor governing people's actions. The shamanic tradition has been accepted in the area as it provides solutions to disaster-related uncertainties. Shamanic tradition proposes eco-friendly, nature-based solutions, but they are insufficient to address the challenges of climate change. Formal institutions find it challenging to intervene to protect against disasters, as seen in the case study of Malana. When these beliefs align with local traditions, their impact is even more profound.

Across different cultures, belief systems are a crucial part of society and play a significant role in how people perceive and respond to such uncertainties. The chapter highlights various ways people address their uncertainties with different examples. While some people may abandon their beliefs during periods of uncertainty, others view religion as a beacon of hope for disaster recovery. They see it as a source of comfort and optimism that things will eventually improve. It is vital to establish an interface between formal and informal institutions during times of disaster and uncertainty.

To effectively manage and mitigate the impact of disasters, it is essential to consider cultural and religious factors. This involves collaborating closely with religious leaders, understanding local customs and beliefs, and devising plans that align with them. By building strong relationships with religious communities, trust can be established, and cooperation can be fostered in times of crisis. Clear and open communication, active community involvement, and education are crucial to successfully managing disasters in diverse religious settings.

Acknowledgements The paper submitted is a part of my master's thesis, submitted under the guidance of Prof. Dr. Jacquleen Joseph, Dean at Jamshedji Tata School of Disaster Studies. The study was conducted from 2020 to 2022. I want to acknowledge that as a compulsory step in the thesis submission process, this paper has been uploaded to the e-thesis portal of the University, ensuring its accessibility and dissemination within the academic community.

References

AIDMI (2018) Understanding uncertainty: views from Kachchh, Mumbai and Sundarbans. https://www.preventionweb.net/publication/understanding-uncertainty-views-kachchh-mumbai-and-sundarbans

Avery C, Zemsky P (1998) Multidimensional uncertainty and herd behavior in financial markets. Am Econ Rev 88(4). https://people.hec.edu/lovo/wp-content/uploads/sites/28/2019/03/Multidimensional-Uncertainty-and-Herd-Behavior-in-Financial-Markets.pdf

Beck U, Ritter M (2010) Risk society: towards a new modernity. SAGE Publications, Los Angeles

Bentzen J (2019) Acts of god? religiosity and natural disasters across subnational world districts*. Econ J 129(622):2295–2321. https://doi.org/10.1093/ej/uez008

Bloom N, Floetotto M, Jaimovich N, Saporta-Eksten I, Terry SJ (2018) Really uncertain business cycles. Econometrica 86(3):1031–1065. https://doi.org/10.3982/ecta10927

Brännmark J (2018) Contested institutional facts. Erkenntnis 84(5):1047–1064. https://doi.org/10.1007/s10670-018-9994-7

Chotkowski J (2010) Market institutions and transaction costs - key concepts of new institutional economics. Ann Agric Sci G Ser Agric Econ 97. http://yadda.icm.edu.pl/yadda/element/bwmeta1.element.ekon-element-000171285379

Chunhabunyatip P, Sasaki N, Grünbühel C, Kuwornu J, Tsusaka T (2018) Influence of indigenous spiritual beliefs on natural resource management and ecological conservation in Thailand. Sustainability 10(8):2842. https://doi.org/10.3390/su10082842

Douglas M (1966) Purity and danger: an analysis of concept of pollution and Taboo. Routledge, London

Hackett C, Grim BJ (2012) (rep) The global religious landscape a report on the size and distribution of the world's major religious groups as of 2010

Hebb DO (1954) The problem of consciousness and introspection

Hendriks B (2010) Urban livelihoods, institutions and inclusive governance in nairobi: "spaces" and their impacts on quality of life. Influence Polit Rights. https://doi.org/10.5117/9789056296360

Jovanovic B, Ma S (2020) Uncertainty and growth. Disasters. https://doi.org/10.3386/w28024

Liberatore F, Pizarro C, de Blas CS, Ortuño MT, Vitoriano B (2013) Uncertainty in humanitarian logistics for disaster management. A review. Decis Aid Models Disast Manage Emerg 45–74. https://doi.org/10.2991/978-94-91216-74-9_3

Misiura G, Rozkwitalska M (2019) Institutions, trust and the high level of innovativeness in the Scandinavian countries. WSB J Bus Finan 53(1):44–54. https://doi.org/10.2478/wsbjbf-2019-0005

Murmu C, Pramanik R (2018) Shamanistic beliefs and practices: emerging trends towards a new horizon. https://www.researchgate.net/publication/335839879_Shamanistic_Beliefs_and_Practices_Emerging_Trends_towards_a_New_Horizon/link/5d7f861792851c22d5dbb55e/download

Parthasarathy D (2018) Inequality, uncertainty, and vulnerability: rethinking governance from a disaster justice perspective. Environ Plan E: Nat Space 1(3):422–442. https://doi.org/10.1177/2514848618802554

Pargament KI, Koenig HG, Tarakeshwar N, Hahn J (2004) Religious coping methods as predictors of psychological, physical and spiritual outcomes among medically ill elderly patients: a two-year longitudinal study. J Health Psychol 9(6):713–730. https://doi.org/10.1177/1359105304045366

Regad M, Silva S (2020) The role of beliefs and religion in disaster. In: Culture and risk. https://cultureanddisaster.eu/handbook/1.2-role-religion-beliefs

Rocha J, Fernandez IB, Xia W, Gudi A (2009) Dealing with task uncertainty in disaster management: the role of knowledge sharing for exploration and exploitation. In: Americas Conference on Information Systems (AMCIS). https://aisel.aisnet.org/cgi/viewcontent.cgi?article=1713&context=amcis2009

Schein E (ed) (2012) What is culture? In: Sociology of organizations. Pine Forge Press, USA

Schnarr A, Mertz M (2022) States of uncertainty, risk–benefit assessment and early clinical research: a conceptual investigation. Sci Eng Ethics 28(6). https://doi.org/10.1007/s11948-022-00418-w

Sherry J, Curtis A (2017) At the intersection of disaster risk and religion: interpretations and responses to the threat of Tsho Rolpa Glacial Lake. Environ Hazards 16(4):314–329. https://doi.org/10.1080/17477891.2017.1298983

GPSR Compliance

The European Union's (EU) General Product Safety Regulation (GPSR) is a set of rules that requires consumer products to be safe and our obligations to ensure this.

If you have any concerns about our products, you can contact us on ProductSafety@springernature.com

In case Publisher is established outside the EU, the EU authorized representative is:

Springer Nature Customer Service Center GmbH
Europaplatz 3
69115 Heidelberg, Germany

The manufacturer's authorised representative in the EU is Springer Nature Customer Service Centre GmbH, Europaplatz 3, 69115 Heidelberg, Germany. If you have any concerns regarding our products, please contact ProductSafety@springernature.com

Printed and bound by CPI Group (UK) Ltd, Croydon, CR0 4YY
01/10/2026
02225898-0016